Writing and Managing Winning Technical Proposals

Tim Whalen

Holbrook & Kellogg Inc.
1964 Gallows Road, Vienna, Virginia 22182
Phone (703) 506-0600   Fax (703) 506-1948

Writing and Managing Winning Technical Proposals

Revised Third Edition

Tim Whalen

Holbrook & Kellogg Inc.
1964 Gallows Road, Vienna, Virginia 22182
Phone (703) 506-0600   Fax (703) 506-1948

About the Author

Tim Whalen has over twenty-five years in proposal management and proposal writing with firms such as Frank E. Basil, GTE Spacenet, Systemhouse, Vitro, Boeing, Williams Brothers and others. He holds Bachelor's and Master's degrees from The University of Tulsa. He began his career as a missile launch/classified codes officer in the U.S. Air Force, Strategic Air Command. Mr. Whalen has been an instructor in Business Communication at The University of Tulsa, and taught the special proposal stem in MIT's summer program, Communicating Technical Information. His prior works include the IEEE's comprehensive collection of papers, *How to Create and Present Successful Government Proposals*, co-edited with James Hill; *Writing and Managing Winning Technical Proposals, 1st ed.* and *2nd ed.*; and *Proposals and Feasibility Studies*, plus numerous articles in engineering and technical journals.

Holbrook & Kellogg Inc.
1964 Gallows Road
Vienna, Virginia 22182-3814
(703) 506-0600

© 1994, 1996 by Holbrook & Kellogg Inc.
All rights reserved. No part of this book may be reproduced in any form or media without the written permission of the publisher, except for brief quotations in review articles.

Printed in the United States of America.

Writing and Managing Winning Technical Proposals
HP 4

ISBN 1-56726-033-0

Table of Contents

Chapter 1—Introduction

1.1	Interdependence of Marketing and Proposals	1-1
1.2	The Competition is Potent	1-2
1.3	Proposal Organizations	1-3
1.4	Prewriting	1-3
1.5	Writing Strategies	1-3
1.6	Project People Versus Proposal People	1-3
1.7	Continuous Improvement	1-4
	Source Selection Letters	1-5
	Red Team Final Remarks	1-5
	Independent Analysis	1-5

Chapter 2—Proposal Preparation and Marketing Information

2.1	Marketplace and Management	2-1
	Long Range Marketing	2-3
	Near-Term (Pre-RFP) Marketing	2-3
	Marketing During Period of RFP & Evaluation	2-4
	Postaward, Pre-performance Marketing	2-5
	Marketing of Ongoing Projects	2-5
	A Review	2-7
2.2	History, Experience, and Disciplines	2-8
2.3	Technical Marketing Specialists	2-11
2.4	Technical Marketing Managers	2-12
2.5	Critical Marketing Intelligence and Strategies—RFP Preparation and Influence	2-13
	2.5.1 Renewal: Writing the Incumbent Proposal	2-14
	The RFP Process	2-15
	Holding On	2-15
	The Preproposal Process	2-16
2.6	The Management—Marketing Plan	2-21
2.7	Contract Marketing Archives	2-22
	2.7.1 Proposals in the Aftermath: Postmortems	2-23
2.8	Secure Proposal Operations	2-28
	2.8.1 Security in Proposal Operations	2-30
	Primary Security Features	2-30
	Computer Security Features	2-31

© 1996 by Holbrook & Kellogg, Inc.

		Understanding the Competition	2-32
		An Example—The Walk Around	2-33
2.9	Feasibility Studies as Marketing Tools		2-34
		Targeting the Study	2-35
		The Possible Disadvantages	2-37

Chapter 3—Sizing up the Competition

3.1	Data Base—Public and Private		3-1
		The Freedom of Information Act	3-2
		Public Information	3-2
		Private Information	3-3
	3.1.1	Congress and Congressional Agencies	3-5
		Lobbying—A Regulated Industry	3-5
		Tracking	3-6
		Counterparts	3-6
	3.1.2	Audit Agencies	3-7
		The General Accounting Office	3-8
		The Defense Contract Audit Agency	3-9
		The Inspector General	3-10
		Performance—The Key To Success	3-10
	3.1.3	Public Sources of Data for Assessing the Opposition	3-11
	3.1.4	Private Sources of Data for Assessing the Competition	3-13
3.2	Capability and Qualifications Statements		3-14
3.3	Conferences, Associations, Professional Societies and Technical Journals		3-14
3.4	Independent Research and Development (IR&D) Projects		3-15
3.5	Feasibility Studies—Preparation or Interception?		3-16
	3.5.1	Executive Summaries and Feasibility Studies	3-18
	3.5.2	Technical Appendices to Feasibility Studies	3-19
3.6	Preliminary RFPs		3-20
	3.6.1	Advance Work on the RFP	3-21
	3.6.2	Simulated RFPs	3-21
		Rating the Client's Priorities	3-21
		White Papers and Drafts	3-22
		Goals of the Simulated RFP	3-23

3.7	Go—No Go Strategies	3-24
	The Competitor's Viewpoint	3-25
	The Ten-Step Approach	3-26
3.8	Adjustments to Stay Competitive	3-30
3.9	Summary	3-31

Chapter 4—Proposal Development Organizations

4.1	Proposal Managers and Proposal Directives	4-4
4.2	A Dedicated, Permanent Proposal Development Organization	4-9
	4.2.1 PDO Manpower	4-10
4.3	PDO Alternatives for Small and Medium-sized Firms (*AD HOC* Organizations)	4-12
4.4	Least Favorable Choice	4-12
4.5	Proposal Costs: Meeting the Milestones On Time and On Budget	4-13
	4.5.1 Annual Plan	4-14
	4.5.2 Cost Logic for Solicited Proposals	4-15
	4.5.3 Cost Logic for Feasibility Studies	4-19
4.6	Final Considerations	4-20

Chapter 5—Proposal Prewriting Activities

5.1	Standard Ingredients of Proposals	5-2
	5.1.1 A Typical Proposal Outline	5-4
	5.1.2 Typical Proposal Categories	5-5
	5.1.3 Summary of Ongoing Activities	5-13
5.2	Source Selection Boards—The Selection Process	5-15
5.3	Prewriting and Marketing	5-16
5.4	Developing an In-House Preparation Course	5-17
	5.4.1 Scope of the Course	5-19
	5.4.2 Finding the Right Sponsor and Authority	5-19
	5.4.3 Useful Publications	5-20
	5.4.4 The Final Ingredient	5-20
	5.4.5 Timing	5-21
5.5	Prewriting Results	5-22

Chapter 6—Where RFPs Come From and Why They Must Be Obeyed

	Funding the RFP	6-1
	Preparing the RFP	6-3
	Knowing the Client's Needs	6-4
6.1	RFP Matrix and Writing Plan	6-7
6.2	RFP Compliance Matrix Checklists	6-12
6.3	Improved Proposal Writing: Unity, Coherence, and Emphasis	6-12
	Be Persuasive	6-12
	Be Logical	6-24
	Control the Presentation	6-25
	Review the Presentation	6-25
	Make a Final Evaluation and Revise Accordingly	6-27
	L and M Coordinates	6-27
6.4	Writing Past Performance & Similar Experience: the Proactive Articulation of Corporate Credentials	6-30
	The Citations Are Limited by the RFP	6-32
	Using the YES/NO Checklist	6-34
	Preparing Past Performance Sections	6-35
	Past Performance Implications Are Throughout the RFP	6-36
	Implications of the Increased Emphasis on Past Performance	6-43
	Government Adjustments	6-44
	Conclusion	6-48
6.5	Writing the Resume Section	6-48
6.5.1	Understanding How Clients Score and Evaluate Resumes	6-49
6.5.2	Avoiding Bait-and-Switch Accusations	6-50
6.5.3	Writing the Technical/Business Resume	6-51
	The Telegraphic Resume	6-53
	The Scholarly Resume	6-53
	The Categorizing Resume	6-55
	The Job Application Resume	6-56
	The Balanced Resume	6-57
	Summary of Technical and Business for Proposals	6-65

© 1996 by Holbrook & Kellogg, Inc.

Chapter 7—The Proposal Manager

7.1	The Proposal Manager's Directives	7-4
	Kick-Off Meeting Directive	7-4
	Win Strategy Marketing Information and Theme Development Directive	7-5
	Blue Team Directive	7-6
	Red Team or Midterm Corrective Action Directive	7-6
	"Time to Finish" Directive	7-7
	Completion Directive	7-7
7.2	The Proposal Manager's Directives And Quality Controls	7-7
7.3	Authority And Budget	7-8

Chapter 8—The Proposal Manager's Tools

8.1	Writing Strategies	8-5
8.2	The Oh-Oh Principle (Negative Information)	8-5
8.3	Competitors Have Faults, Too (The Ghost Story Principle)	8-6
8.4	Discriminators	8-8
	Adverse Iteration	8-8
	Ghosting	8-10
	Typical Discriminators	8-11
8.5	The Red Team	8-13

Chapter 9—The Storyboard, STOP and Other Writing Techniques

9.1	Writing Team And Storyboard	9-3
9.2	Other Writing Strategies	9-10
9.3	The Hughes Aircraft Method	9-11
9.4	The Martin Marietta Method	9-11
9.5	Typical Sheets For STOP	9-11
9.6	Typical Sheets For STEP	9-12
9.7	Arts And Graphics In Storyboarding	9-13
	Uniformity	9-14
	Style	9-16
	Quantity	9-16
	Placement	9-16

© 1996 by Holbrook & Kellogg, Inc.

		PERT	9-17
		Gantt	9-17
9.8	Using Appendices		9-21
9.9	Lengthy Items		9-23
9.10	Tangents		9-24
9.11	Stretching Page Limitations		9-24
9.12	The Volumes In Summary		9-26
	9.12.1	Executive Summary	9-26
	9.12.2	The Technical Proposal	9-28
	9.12.3	The Management Proposal	9-28
	9.12.4	The Cost Proposal	9-29
	9.12.5	Scopes of Work, Specifications, and Sample Work	9-30
	9.12.6	Tailoring Resumes for Proposals	9-30

Chapter 10—The Final Product

10.1	STOP Writing		10-1
		Developing the Thesis	10-1
		Developing the STOP Profile	10-2
		Drafting the STOP Unit	10-4
10.2	STOP Editing		10-5
		The Initial Review	10-6
		The Final Review	10-8
10.3	Production, Layout, And Publication		10-10
		Typesetting	10-11
		Graphic Art	10-11
		Proofreading	10-11
		Layout	10-12
		Page Make-up	10-12
		Printing	10-13
		Binding	10-13
10.4	Quality Control Checks		10-15
10.5	Packaging And Delivery		10-16
		Packing an Unclassified, Nonsensitive Document	10-17
		Packing a Classified or Sensitive Document	10-18
		Delivery	10-18
10.6	Archiving, Cleaning Up, And Debriefing		10-19

© 1996 by Holbrook & Kellogg, Inc.

Chapter 11—Proposals in Retrospect

11.1	The Teams—White, Blue, Red, And Gray	11-1
11.2	One Dimension—The White (Writing) Team	11-3
11.3	Two Dimensions—The Blue Team	11-5
11.4	Three Dimensions—The Red Team (The Devil's Advocate)	11-6
11.5	The Fourth And Final Step—The Gray Team (Win-Loss Assessment)	11-9
	Major Proposals	11-11
	Minor Proposals	11-11
	Prognosis—Major Efforts	11-11
	Prognosis—Minor Efforts	11-12
	Next Year's Synthesis-Full-Cycle Planning	11-12

List of Figures

2.1	The Continuous Marketing Cycle	2-2
2.2	Preproposal Writing Activities	2-16
2.3	Proposal Postmortem Summary Sheet	2-25
2.4	Archival Proposal Data	2-26
2.5	Security Flow for Proposal Development	2-29
2.6	Initiation of a Feasibility Study	2-36
4.1	Proposal Development Organization Interfaces	4-3
4.2	Proposal Directive Memo and Schedule	4-5
4.3	Proposal Flow Chart	4-16
4.4	Proposal Cost and Schedule Plan	4-17
5.1	Summary of Essential Proposal Criteria	5-13
5.2	In-House Proposal Preparation Courses	5-18
6.1	Proposal Operations	6-6
6.2	Typical Proposal Staff Structure	6-9
6.3	L and M Coordinates—Emphasis and Economy	6-29
6.4	Get a Jump on the Evaluators—Scrutinize All Data for Exact Past Performance Matches	6-42
6.5	Telegraphic Resume	6-59
6.6	Scholarly Resume	6-61
6.7	Categorizing Resume	6-62
6.8	Job Application Resume	6-63
6.9	Balanced Resume	6-64
7.1	Typical Proposal Time Lines	7-2
9.1	STOP Symbol	9-3
9.2	STOP Storyboard Sheets—	
	Sheet (a)	9-5
	Sheet (b)	9-6
	Sheet (c)	9-7
9.3	"Whole before Parts"	9-9

© 1996 by Holbrook & Kellogg, Inc.

9.4	IEEE and IBM Standard Symbols	9-15
9.5	PERT Chart for Proposals	9-19
9.6	Typical Gantt Chart for Software Development	9-20
9.7	Balanced Resume	9-32
9.8	Ideal Proposal Room	9-34
11.1	The Four Teams for STOP Proposal Writing	11-3
11.2	Improving the Proposal	11-7
11.3	The Red Team's Review Reflects Its Knowledge of the Ratio of Successful Proposals to Overall Marketing Actions	11-8

List of Tables

2.1	Documented Incumbent Advantages versus Adversaries	2-17
3.1	Evaluation Factors for RFPs	3-28
3.2	Adjusting for Errors	3-31
5.1	War Game Scenario for Typical Client's RFP Outline-Estimate	5-4
5.2	Categories of Essential Information for Proposals from	
	U.S. Industry	5-8
	U.S. Army	5-9
	U.S. Navy	5-10
	U.S. Air Force	5-11
	U.S. DOE	5-12
6.1	Preliminary Table of Contents	6-10
6.2	Sample RFP Checklist	6-14
6.3	Detailed Proposal Evaluation Checklists for Authors and Reviewers	6-15
8.1	Using Discriminators	8-9
9.1	Quality Assurance Approach—Exploded Version	9-8
9.2	Sample Proposal and Two Detailed Sections	9-25

© 1996 by Holbrook & Kellogg, Inc.

CHAPTER 1

Introduction

1.1 INTERDEPENDENCE OF MARKETING AND PROPOSALS

No two better allies exist in the new business arena than marketing teams and proposal teams. These two teams are involved in a unique, symbiotic relationship, in which both prosper or suffer according to the acceptability of a stack of pages called the *technical proposal*.

The proposal's acceptability to clients is the key to winning, and it takes strong marketing and strong proposal preparation techniques to deliver the winning product. Marketing and proposal preparation have to be in perfect harmony—no fits, false starts, or imprecise estimates—in order to write that special set of words that will be awarded a new contract. The interdependence of proposal marketing strategy and marketing is crucial. If the marketing strategy is poorly conceived and poorly written, we cannot develop a persuasive proposal strategy. If the marketing plan is weak or unconvincing, our proposal will not capture the client's imagination (or the contract!).

However, despite this vital interdependence, marketing and proposal groups are separate entities in many companies. This book aims to break down the barriers that prevent these groups from working together and producing a winning proposal. Engineers, managers, marketers, and technical or proposal specialists who use this book will see that parts of it pertain to the entire proposal production process, while other parts pertain to only certain aspects of the process. Savvy marketing and

© 1996 by Holbrook & Kellogg, Inc.

polished writing unite to create an unbeatable proposal; one without the other cannot withstand the pressure generated by competent, aggressive competitors.

1.2 THE COMPETITION IS POTENT

Each year in planning sessions, many firms convince themselves that the next 12 months are going to be different—this time, the competition is going to be routinely defeated. Individual managers exude the same unfounded confidence, glowing with the assumption that "the other guy" is ill-informed, ill-prepared, and working in the dark.

On the contrary, if competitors have been winning much of the time, then statistically they can be expected to continue to do so. Chapters 2 and 3 describe how to evaluate and prepare for the competitors, before they research and size up our own firms.

Overly optimistic thinking usually produces a "halo effect" around the team, inducing it to think it will defeat all opponents. This state of mind would be like that of a diplomat who expects the enemy's diplomats to be blind, deaf, dumb, and crippled. The bright, clever diplomat knows that although negotiations are long, hard, and disappointing, they are also invigorating, energetic, and rewarding. Each situation offers fresh approaches to old problems.

We may also compare marketers and proposal coordinators to athletes. Strength, stamina, endurance, training, natural ability, and coaching all make a difference. In team sports—baseball, basketball, soccer, hockey, or football—it is the team that wins; individuals fit into the team's program. Uncooperative rogues are not tolerated by management or the other players. Although the performances of a few players stand out, individual efforts blend into a concerted group effort. A poor athletic team does not know the other team's weaknesses or strong points, nor how to modify its strategy for every opposing team. Also, a poor team does not realize that some of those "other guys" play dirty. Winning teams do not go into the competition without knowing the rules, the field, "the other guy," and how to put it all together.

1.3 PROPOSAL ORGANIZATIONS

There are as many ways to write proposals as there are companies, but many of these ways are crude, ineffective, expensive, and bound to repeat many previous mistakes. The permanent proposal coordinating entity is discussed at length in Chapter 4, and is used throughout in contrast to the pandemonium many engineers and managers regard as "typical" proposals. We cannot stress enough that organized proposals are probably twice as likely to succeed as those "fly by the seat of the pants" exercises, which Hy Silver so rightly dubbed "stampedes."

Engineers in private practice and smaller firms can tailor these suggestions for proposal development organization down to a size and scope that is manageable, economic, and effective.

1.4 PREWRITING

As Chapter 2 emphasizes, a lot of preparation and organization goes into the proposal before we begin to write the first draft. We need to evaluate the needs of our clients, research the topic, understand the standards and specifications, and finalize contracts. This investment in time in the early stages of proposal production ensures the delivery of a high-quality product.

1.5 WRITING STRATEGIES

Chapters 6, 7, 8, 9, and 10 explain general writing and storyboarding techniques, which are so essential for success in today's marketplace. We enclose checklists in those chapters that you can use to audit your own proposals.

Again, smaller firms and engineers in private practice can scale down the storyboard and other strategic writing plans to make them manageable.

1.6 PROJECT PEOPLE *VERSUS* PROPOSAL PEOPLE

In terms of personnel, there is a drastic difference between people who work on proposals and people who work on projects. Here are some characteristics:

© 1996 by Holbrook & Kellogg, Inc.

Factor	Project People	Proposal People
Work	9 a.m. - 5 p.m. period, no weekends	8 a.m. - 6 p.m. or later, weekends as needed
Attitude	Unenthusiastic, modest	Energetic, assertive
Goal	Finish the job	Produce a winner

There is no room for a halfhearted proposal member or team. The goal is to win, actually to defeat every other player and take the award. Long hours and hard work are typical. During the 30 to 40 days required to produce a proposal, project people have to become proposal people, or at least imitate proposal people. The entire group must ensure that every member has a winning attitude.

1.7 CONTINUOUS IMPROVEMENT

Your goal throughout the year is one of continuous improvement. Just as you practice and improve your skills with this book, your comprehension and knowledge of recent proposals similarly adds to your skill sets. As a preview of the book, and as a checklist for continuous improvement year after year, the following points of emphasis are provided.

While reviewing all of last year's proposals, you should identify six winning proposals and six losing proposals. You still have confidence in the approaches and tactics you used to write the six proposals that were unsuccessful. However, you do not want to repeat any of last year's mistakes this year. What can you do to assure an assimilation of your good traits and the elimination of traits that were not successful?

There are a variety of product enhancements to institute at the beginning of each marketing cycle. The first is a complete validation of marketing data with multiple, independent sources. In other words, of the six nonsuccessful documents, how many should not have been written due to a lack of valid business intelligence? Here are some items to review side-by-side from a common vantage point, and not in isolation:

Source Selection Letters

What did they criticize? Is there a pattern of "No Thanks" remarks in letters from several agencies, or from the same agency? The source selection authority (SSA) and source selection committee reduce all their findings to writing, so a careful reader can discern even the most diplomatic criticism, given a thorough reading. Do not overlook praise from the SSA as genuine responses to work well done.

Red Team Final Remarks

When your proposals were completed after the Red Team review, what specifically did the Red Team Captain tell you to watch out for? Did the captain mention any remaining weaknesses to overcome by BAFO or in negotiation? What, systematically, did the Red Team repeat about the unsuccessful proposals? What did they emphasize about your winning proposals? Correlate the relationships you discover in the Executive Summaries, Technical Approaches, Business/Management Plans, and Cost.

Independent Analysis

The third and most rigorous investigation of your trends in winning and losing proposals can be made using outside independent experts. In one proven test, two or more persons with credentials from other branches of your firm, from business allies, or from consultants are given a batch of proposals and the respective Requests for Proposals (RFP), but without the benefit of knowing which documents won and which did not. In this acid test of proposals and consultants, you will glean a fresh perspective of your proposals.

The outsiders do not know which proposals are now bringing you income, and which ones only cost you overhead money. Thus, you are in a unique position to judge the findings. "They're all winners" is a grossly over-optimistic opinion, avoiding any substantial criticism for fear of offending management. "They're all faulty" is a similar finding; overcritical, overzealous, and for you, obviously, not a true statement. It is likely that the better opinions will make statements such as:

- The Executive Summary identified many special discriminators that made your firm look not only well qualified, but the best qualified, a solo choice, the one unique proposer; even your transmittal letter contributed to your uniqueness, dedication, and competence.

- The Management Plan was well tailored to the client's needs from our G-2, from agency briefings, and from the RFP.

- The Technical Proposal not only met all RFP criteria, but showed *how* and *how well* you will perform. You showed a clear sense of direction, complete understanding, and important improvisations that will make the client appreciate you. Your graphs and charts were informative, proving to the client that you have a mastery of their problems.

- Costs were well documented; tracked to the Statement of Work (SOW); and were lean, mean and aggressive, giving options for the client to route funds if and when the project grows in those directions.

- Resumes showed a genuine sense of compatibility, harmony, and dedication. Every person fit into the SOW as a dedicated professional; educated, experienced, and proven in his or her role. The project manager has an exemplary track record for success in this and related work; all supervisors reflect sub-sets of the project manager's overall expertise. Moreover, the resumes are easy to read, brief, and highlight project-essential skills.

- Appendices, Sample Problems, or Sample Task Orders showed a high degree of accuracy in projecting skills needed and quantities/durations/economies of scale. Professionalism was evident even in the least of the sample deliverables, giving the client confidence that you are the one right choice.

Turning now to the not-so-flattering trends discovered among your proposals, here is a list of unpleasant findings that are typical of a review of a year's outputs of proposals, including some winning and some losing efforts:

- The Executive Summary could have gone ahead of any proposal. There was no evidence that you knew what the priorities were. Your choice of subcontractors and their respective roles was imprecise in the extreme—how they work together is a mystery, and there seems to be a lot of overlap. Consequently, costs may be too high.

- The Management Plan was generic and not remarkable for this opportunity; there were loose ends and terminology not related to the RFP.

- The Technical Proposal followed the RFP, reiterating the SOW, but with no genuine innovations, approaches, economies of scale, or discriminators; it was an empty shell seeking substance. The illustrations were boilerplate, drawn from stock standards or specifications, and lent no further credibility to your claims of excellence.

- Costs were jumbled, lumped together, and hard to understand. There is a general sense that something is being concealed.

- Resumes are problematic. They are in different fonts and styles. It is difficult to find project-related significance. Some resumes appear out-of-date and arbitrarily included. The person appearing as project manager has a fine career, but how is it related to the SOW? Supervisors and technical persons appear, from their own credentials, to be deeply committed to other programs; how will they suddenly "jump ship" and appear on the new work?

- Appendices, Sample Problems, or Sample Tasks are apparently the work of others; they do not have the same layout, texture, or approach as the Technical Proposal. Some appear hurriedly prepared, full of gaps or assumptions, and not the kind of work that would be acceptable to the client on a day-to-day basis. If the proposal is in production and/or manufacturing, was the sample or test item 100 percent compatible to the client's guidelines? If not, the sample of work is simply put into the category of "Unsatisfactory."

There should be a variety and range of opinions, and some surprises in what the selected readers term "good" and "bad." This is a healthy controversy, opening discussions of items you may regard as static—like resumes—and highlighting mistakes not to make again. The firm has at its fingertips the key information and knows which proposals really won, and suspicions of why the others lost. Hence, you retain control of all the results of this test.

If a firm has the time and the funds to do two analyses per year of its proposals, then the investment is prudent. The first review is best scheduled in January and the second in July. If only a single review can be done, the January session will have to suffice to correct all of last year's perceived shortcomings.

© 1996 by Holbrook & Kellogg, Inc.

The three methods, shown above, are all productive ways for improving your own proposal process. The third method, however, by uninvolved, independent outsiders is recommended as the most enlightening approach, validating the good and bad traits of last year's work in an unbiased, yet controlled, atmosphere.

CHAPTER 2

Proposal Preparation and Marketing Information

2.1 MARKETPLACE AND MANAGEMENT

The importance of marketing intelligence in proposal preparation has been vastly underrated and underestimated. In fact, each year many losing proposals have excellent technical approaches, excellent management concepts, and excellent cost factors. So why were so many fine products rejected?

Marketing strategy determines the winners and the losers. It identifies the clients who would be most interested in the proposal. Marketing is the criterion by which the client judges the proposal firm. Without an excellent marketing team to guide the proposal preparation group, the proposal is likely to get lost in the marketplace, devoid of the necessary opinions, attitudes, and facts that are sorely needed to win. The key is to get in front of the client—to be physically present to brief, visit, and solve problems—and be there as often as the client allows. Between marketing and proposal groups, there should be a symbiosis and shared pool of information about what is developing in the marketplace, and discussion about what to do with that information. Proposals *must* benefit from every shred of marketing information.

In the continuous marketing cycle for proposal preparation, which we itemize in Figure 2.1, there are typically five continuous areas of concentration. These are:

- Long-range marketing

- Near-term marketing

- Marketing during period of RFP evaluation

- Postaward, pre-performance marketing

- Marketing of ongoing projects

Successful contractors market clients continuously, year-round, concentrating on being completely prepared for each proposal, with emphasis on these five special time frames:

- *Long Range Marketing* – market specific clusters of related clients, as much as 1-2 years ahead of an RFP, offering the clients concepts to shape the marketplace with innovations that each bear your own trademark.

- *Near Term (Pre-RFP) Marketing* – market the client with pieces of the technical approach, anticipating what the client will and will not validate as suited to their needs, ahead of any RFP mandates; intensify telling the client how good your team is with informal, social customer contacts as well as formal client briefings.

- *The Period When the RFP Is Out, Proposals are Prepared and Evaluated* – market the client socially and get feed-back; influence and upgrade the proposal at every opportunity via each RFP Amendment and the questions made as Government Deficiency/Correction Reports (CRs, DRs).

- *Post-Award, Pre-Performance Marketing* – negotiate avenues of innovation, preplanned areas of growth, and areas in which to excel early, setting the pace for the on-going project.

- *On-Going Projects* – market all current clients, especially the contracting officer and contracting officer's technical representative and monitors to demonstrate areas where you are and will make discriminating progress.

Figure 2.1
The Continuous Marketing Cycle

© 1996 by Holbrook & Kellogg, Inc.

Long-Range Marketing

Perhaps the most nebulous and time-consuming approach, long-range marketing identifies the targets that will be pertinent in three to five years or more. It develops strategies to position the firm in mainstream technologies or services where the marketplace will logically go. Independent research and development (IR&D) projects can then be budgeted and planned for larger firms, and proportional investments can be made by smaller firms. Marketers with an astute long-range plan make numerous trips and visits to meet potential clients, members of the source selection board, and influential members of Congress, thus influencing them years in advance of actual proposal preparation.

Near-Term (Pre-RFP) Marketing

Perhaps the most intense and critical marketing is near-term marketing, the corollary to long-range marketing. In anticipation of a promising solicitation, marketing professionals must intensify their activities in very specific ways, including the following:

- They must increase the frequency of visits, trips, telephone calls, and official inquiries to the client at the client's headquarters and procuring office. The goal is to keep the firm in the center of the client's thought patterns.

- They must broaden their base of knowledge via satellite offices, subcontractors, vendors, consultants, and political sources. The goal is to absorb the widest data base of the client's overall pattern of behavior, at least 60 to 90 days before the RFP.

- They must employ technical persuasion methods such as suggestions, recommended courses of action, specification preparation, and writing a *scope of work*. The goal is to affect the client's solicitation so the firm can respond uniquely, thus positioning itself to defeat the competition.

- They must submit papers, research reports, conference proceedings, deliverable reports, manuals, and successful development and project specifications to their clients. The goal is to emphasize certain areas of the firm's expertise, and establish the firm as the technological and managerial front runner.

© 1996 by Holbrook & Kellogg, Inc.

As the date of the RFP approaches, the doors that were typically open at the client's offices begin to close. In other words, it is essential to have groomed the clients for receiving the proposal for two to three months prior to the client's releasing the RFP. Near-term marketing ensures that the firm's ideas are ingrained in the client's mind.

After an RFP is released, doors do close on marketing personnel, and calls are not returned. This blackout lasts for 30 to 90 days while the client's proposal evaluation team, or *source selection board (SSB)*, does its job. However, successful near-term marketing will already have influenced the client in favor of the firm's proposal.

In summary, near-term marketing is intense, and brings home to the client the firm's immediate usefulness and excellence, qualities first established with less intense long-range marketing accomplished years before.

Marketing During Period of RFP and Evaluation

Although a client officially evaluates proposals only during the period of RFP, in an atmosphere free from influence and persuasion, not all the client's doors are shut. Source selection boards consist of second level contributors who do not sit on the board, yet are free to move among its members.

While the RFP is out, and later while proposals are being evaluated, no offerer may talk with the SSB. Therefore, long-range and near-term marketing must identify friendly parties on this secondary tier of the SSB, and continue to place persuasive materials in their hands.

This time frame is critical because the client is in the process of choosing the winner. Old enticements can die fast if not reinforced with the latest persuasive data. The marketer should submit a persuasive piece of information *while the proposals are under consideration.* Examples of information that provoke the most successful results with an SSB are:

- News bulletins stating that the test vehicle (or prototype piece of equipment) actually works under field conditions;

- A technical breakthrough was made that is a key to the approach;

- An important manager has been added to enhance the firm's posture;

- The Administration, a governor, a senator, or a representative has endorsed the proposal as *the most advantageous offer overall*; or

- Some part of the competitor's proposal is invalid, erroneous, or has major shortcomings, as demonstrated by a lack of performance in a related, similar project.

Thus, marketing during the period of RFP and evaluation greatly contributes to the success of proposals.

Postaward, Pre-performance Marketing

At this stage, the firm has won, and is in the best graces possible. However, marketing efforts *are not concluded* with the delivery of the proposal to the client's SSB. On the contrary, the *executive summary* and other key elements of the technical and business proposals should be summarized in letters and notes, then applied through the political infrastructure to alert friends and allies that an award has been made and that performance will soon begin.

The goal of this marketing effort is to emphasize employment opportunities, benefits that the community will receive, and to assess *why* the firm won, and how best to exploit the victory. Often, a shrewd negotiating team can add options to the contract, and increase its value, size, and stature. These items added to the proposal may be expressed first as avenues of growth implanted in options to the technical proposal.

Marketing of Ongoing Projects

The last interaction between proposal preparation and marketing is that of ongoing projects. Although they are an obvious source of additional business, ongoing contracts do not receive the kind of attention they deserve. Existing contracts need to receive a lot of marketing attention to defend them from competitors (interlopers) who may try to sway the client in favor of concepts outside the firm's expertise.

Because competent ongoing contracts are the best promotion for new proposals, marketing ongoing projects in terms of contract performance should be among a firm's most significant activities. The reasons are as follows:

- No one else has the contract, the intimate knowledge, or the special relationship to expand the initial scope of work.

- The opposition has been humbled—the competitors have little influence.

- Ordinary progress reports and accomplishments required by the contract can be used as demonstrations of successful performance worthy of expanded or additional work.

- Nothing has the impact of immediate, recent project accomplishments such as deliverable manuals, reports, drawings, and specifications. As appendix materials, these items are lethal to the opposition.

In summary, while ongoing project marketing is quite ordinary, it has a high probability of success, and an inside sales perspective that is unmatched in the competitive arena. Project marketing has the additional benefit of being shouldered mainly by project—that is, billable—employees who prepare the marketplace for professional marketing and proposal development personnel. It is essential that the reader understand that we are discussing *directed marketing* here, not general or undirected marketing.

Each marketing person has his or her precise assignments and is responsible for results—that is, getting in to see the appropriate decisionmaker at the client's office; socializing with them away from the office if possible; learning the policies, politics, and prejudices of the decision-makers; and tabulating what the client expects from the next contractor. Generic or imprecise marketing is far removed from this results-oriented environment. When ever a marketing person answers a management question with "I don't know," the reply from management should be "Find out, promptly." Directed marketing produces management true to the results of the firm's goals; generic marketing can produce an unallied bevy of problematic opportunities.

A Review

The links between proposal preparation and marketing are fundamental to a company's success with new contracts. It is essential that the marketing and proposal development organizations recognize these sources of marketing information as summarized for each of the aforementioned marketing stages.

Long-Range Marketing. No firm has a grip on its future without a credible long-range marketing plan. Electronics firms in 1987 must consider the ATF and LHX projects, and other procurements of the 21st century. The long-range plan sets goals three to five years ahead of today's goals. What are your firm's goals? Where will you be (or want to be) in five years? Where is your marketing plan now leading you?

Near-Term (Pre-RFP) Marketing. The firm that wins the near-term battle often wins, period. Despite a lengthy buildup and the establishment of credibility, some firms lose the job just before the RFP is distributed. It is crucial that the firm score marketing successes in the time allocated to pre-RFP preparations. Nothing can substitute for marketing emphasis at this stage. Pull all the plugs here—make maximum use of persuasion to convince the client of the quality of the firm's offer.

Marketing During Period of RFP and Evaluation. While the proposals are officially being scrutinized, it is prudent to assert the firm's excellence via friendly parties inside and around the SSB. No effort to influence the SSB should be omitted. The key element at this stage is the proposal: it must be thorough, complete, and concise.

Postaward, Pre-performance Marketing. The gap between being awarded the contract and negotiating can contain some interesting events. Although "award" implies 100 percent success, it is rarely so. For the most part, an award implies so much postaward euphoria that nobody realizes how much the firm will have to grow. At this point, proposal development and marketing personnel must step in and outline where the opportunities will be. Prompt exploration of postaward opportunities can be promising. If the firm finds many new employees, it may accommodate all of them by enlarging the contract, thereby making the award more profitable, or last longer.

© 1996 by Holbrook & Kellogg, Inc.

Marketing of Ongoing Projects. The most direct, accurate means of marketing is to promote ongoing projects. Be sure that each project appreciates *its own* performance. The performance of specific tasks will ensure the reputation and credibility of the work described in the RFP. The firm under contract requires the following elements to promote the project properly:

- A project manager acquainted with business development;

- An assistant manager acquainted with how marketing can increase new contracts via expansions, additions, and other opportunities in the immediate geographical area where a project office is located; and

- Regular visits from the proposal development and marketing departments to collect leads, measure progress, and review deliverables that can be used as evidence in future proposals.

When combined in the correct proportions, these five marketing stages cannot be defeated. The interactions between marketing and proposal preparation are inseparable. Without marketing intelligence, proposal activities are useless.

2.2 HISTORY, EXPERIENCE, AND DISCIPLINES

After a firm has taken an introspective look at itself (Section 2.1) a pattern will emerge as to whether or not marketing is synchronized to proposal preparation. The firm must have documentation of the following elements:

- Its contract history, item-by-item, for at least the last 10 years, or 20 years if available;

- The experience, capabilities and qualifications of the firm in its widest definition; and

- The disciplines *needed to perform* any new task envisioned as an area of emphasis—for example, what technical types of people are needed to do the job? If a firm does not have a single nuclear engineer on staff and may need 10 such engineers to pursue certain business, judgment dictates steering away from that business.

© 1996 by Holbrook & Kellogg, Inc.

The firm also needs a proposal library and contract "morgue" of all past and existing work to provide win-loss ratios, and a "road map" of where the firm came from as well as where it is going.

If a firm is in the private sector, and its clients are in the private sector, relatively free from oversight by audit agencies, then marketing is easier than in Federal circles. The firm's marketing staff must be on a first-name basis with these people on the client's staff:

- President and vice president

- Division managers

- Purchasing staff

All these people can provide insight into the client's acquisition plans, so that the firm's marketers can shape a situation into an opportunity for new work. Long-term relationships are important in this arena to establish trust, confidence, and privacy. Reading the *capabilities statement* and *annual report* of the firm cannot substitute for this kind of closeness, and can only be secondary data. In the private sector, however, it is necessary to read and comprehend both of these documents to see the stated condition of the client.

In Federal circles, the problem of marketing intelligence magnifies. Federal agencies like the Departments of Defense and Energy, or the National Aeronautics and Space Administration (NASA), have massive fiscal organizations which predict and prepare budgets several years ahead of enactment. Federal project offices work under a massive oversight (audit) and management structure. Opportunities to serve the client occur at many points and junctions within this network of budget, performance, audit, and management. The question then becomes, "What are our areas of expertise and where do we fit in?"

These are other sources that provide needed data for Federal procurements:

- Members of Congress, and agency officers;

- Lobbyists who present us favorably;

- Congressional reports of the annual budget for each client agency;

- Detailed departmental budgets of agencies and priority (goal) statements;

- Records from Congressional hearings;

- Trade and business magazines like *Aviation Week*, *Business Week*, and *Signal*;

- General Accounting Office (GAO), Inspector General (IG), and Defense Contract Administration Services (DCAS) reports;

- Data about *potential competitors* who already have contracts with clients in whom our firm is interested; and

- Trip reports from marketing on the client's objectives, biases, worries, preferences, and how much funding may be allocated to each sector.

Trip reports are essential to get to the root of the client's concerns, wants, needs, and fears. Budgets of each Federal agency are criticized by GAO, IG, and DCAS staffs each year. Agencies then react to "fix" the critiqued items. Agency personnel welcome opinions from industry to help the agency solve its problems. This interaction results in:

- Contractor's access to the agency;

- Knowledge of what RFP scopes of work may contain;

- Understanding of what approaches the agency likes or dislikes;

- Relationships with client managers, technical staff, and contracts people; and

- Knowledge of how the client perceives the firm, and its competitors.

Trip reports from marketing, which focus on areas of proposal concern, should be available to the proposal development group from a proposal/marketing library. These types of data can and must enter the proposal in order to persuade clients that proposal promises will translate into project performance. In fact, as shown later, much of the proposal can be put together at this point. Thus, the firm's history, experience, and disciplines ensure that information will emerge which will favorably match the competition.

2.3 TECHNICAL MARKETING SPECIALISTS

The main marketing question at many firms is, "Who markets here?" In many aggressive, entrepreneurial firms, every employee is expected to conduct marketing of some kind. This approach provides a tremendous network of information, inquiries, and leads. However, such an approach is usually of mixed quality. Not everybody is capable of marketing.

Amateurish estimates of opportunities are usually overly optimistic and inaccurate, and clutter the channels with useless data. The constant "everybody markets" concept also detracts from project performance (accomplishments and achievements) because managers demand that their professional and technical staff be market-minded. When marketing is conducted by nonmarketing people, it should be supervised by marketing managers, and certainly not done at the expense of project work.

Who, then, markets? By studying technical firms in the Washington, D.C. and Baltimore area in 1987, and comparing that data to information from 1986 about Houston, Denver, and Los Angeles, we see that many successful marketing departments are staffed with engineers. Why does this happen? A technical person is best suited to comprehend the problems of technical clients.

It is no surprise in very small firms, the president is also the chief engineer who may do his own proposal writing. Medium-sized companies usually make marketing the express job of vice-presidents and managers. These technically oriented individuals typically divide their time to cover new business actions. Only larger, established firms can afford full-time marketing and proposal organizations, the staffs of which are unburdened by technical projects. The salaries, benefits, travel, and expenses of these people are paid out of overhead pools of money.

Engineers with a sense for marketing often rise in the corporation because (1) they are technically competent, and (2) the client does not consider them salesmen. A second source of quality marketing personnel can be professional and technical people who are dedicated enough to *learn the specifics of the firm's technology.*

Working together with engineers, these people become excellent marketing representatives.

In short, marketing should be restricted to experienced, trained, marketing individuals with the knowledge to squeeze success out of business development opportunities. The key to writing a winning proposal is to include *relevant, well-researched marketing information* in the proposal development organization. No individual with genuine entrepreneurial skills, however, should be excluded from marketing matters. Entrepreneurs and venture capital individuals fill out the spectrum of business development people. Skilled in their own fields, they can give technical marketing departments support in key areas, but they will take risks that technical people would not. Thus, their influence should be contained in the infrastructure of a technical marketing department.

A final word on proposal development people in marketing. The client will immediately apprehend the nature of visits and concerns of the proposal development staff. These individuals should be kept at arm's length from the client who will regard them as only non-technical sales people.

2.4 TECHNICAL MARKETING MANAGERS

Another facet of marketing besides who is authorized to market is who, exactly, manages marketing in its *entirety*? Some firms have found that while each vice president should have personal marketing goals, a single manager of vice-presidential status should be in charge. A budget should be assigned to support the corporate needs of that manager.

As with marketing specialists, good marketing managers should have certain qualifications:

- The manager should have 10 years of experience as a marketing specialist.

- He or she was preferably an engineer in the mainstream of the company's business activities.

- The manager is highly skilled in proposal preparation, and knows how to incorporate marketing intelligence into the correct parts of the winning technical proposal.

- He or she is personally successful in persuasion, and was instrumental in winning several contracts.

Thus, marketing management is intrinsic to putting proposals in a winning position.

2.5 CRITICAL MARKETING INTELLIGENCE AND STRATEGIES—RFP PREPARATION AND INFLUENCE

At this point, the logical goals of marketing can be expressed as follows:

- To impress the client favorably with the firm's capabilities;

- To negate the offerings and concepts of competing firms;

- To steer the proposal development group toward extensive prewriting of a proposal and an ideal RFP; and

- To add information to the client's concepts (and to the RFP if possible) that will make a particular contract unique to the firm's offerings, and dissimilar to what competitors have to offer.

This last goal is the most vital. If the RFP can be shaped to the firm, then a successful proposal can be done economically with great certainty of a win. However, how is the RFP prepared by marketing personnel? Let us look at a sample case.

An agency must add a new project to its field offices for a particular type of military hardware and software. Congress has allocated large sums to finance *all* activities. A Federal project manager and a small staff have been assigned to solve this acquisition problem. A number of contractors have written, visited, briefed, and submitted unsolicited proposals to this office, but to no avail. Finally, a firm recognizes that the field office has no confidence in either the contractors or themselves. In short, the agency office wants a cradle-to-grave solution, one that is safe, highly technological, and not too expensive. Thus, the client seeks a contractor who can provide hardware and software to meet these requirements.

The firm then prepares a briefing which addresses all the client's anxieties (and goals). It submits white papers that cover the agency's scope of work. Perceiving this firm as the only contractor concerned enough to be truly responsive, the client's field project manager acquiesces to this special help, and directs his contract people to use the contractor's suggestions and concepts in the government's RFP. In this manner, the RFP will contain no surprises and will be structured so that the aggressive firm can readily win.

While trying to woo Federal clients, the marketing group should remember these situations:

- Some other firm may be entrenched as the client's incumbent or favorite. If our competing firm cannot undermine the incumbent, then we should not bother to bid on the RFP.

- If the client is giving all competitors equal consideration, our firm has to do a lot of marketing work—for example, submit white papers and get organized.

From a strategic vantage point, whether there is an incumbent (or favored) contractor creates many prewriting proposal tasks, as shown in Section 2.5.1.

2.5.1 Renewal: Writing the Incumbent Proposal*

Unhappily, at least within facets of the petrochemical industry, there appears to be a nonrenewal trend of crucial contracts covering architect-engineer services, operations-maintenance services, and construction management services. Of the 300 engineering firms listed annually in *The Engineering News Record*, virtually all have felt the pressure of competition to write renewal proposals for their own contracts. A key element to proposal writing in this atmosphere has been to write preproposal materials for continuance—certainly prior to the time when the incumbent may be forced to compete for the work in the contract via a full, formal proposal.

---

* This section is based on an article from *IEEE Transactions on Professional Communication*, Vol. PC-28, No.2, June 1985.

© 1996 by Holbrook & Kellogg, Inc.

In this uncertain and highly competitive arena, past corporate relationships (and track records) are subject to chronic challenges, as well as *ad hominem* critiques. Writers planning incumbent proposals simply cannot depend on a safe and sure road to renewal, or even of an extension of a year or two, and must focus a great deal of energy on writing to gain a competitive edge.

The RFP Process

The most visible sign of the increase of nonrenewal is the all too often exercised privilege of a client to recompete work via the RFP method, a publicly advertised process which calls for willing competitors to come forward with their own technical proposals aimed at accomplishing the client's goals. (*The Commerce Business Daily* and *The Wall Street Journal* are two of the best known places where RFPs are published.) When such an RFP appears at or near the end of an incumbent's agreed upon term of duration, it is a sure sign that the client is dissatisfied and that the incumbent's performance is in doubt.

When the proposal writer comes to this crossroad, it is time to diagnose the situation properly, and to write to remedy the client's perception of what is to blame.

Holding On

At this point, writers of incumbent proposals still hold many of the key elements of renewal within their own grasp—they are still in the drivers' seat, and have the resources and assets to keep the contract under control. What may at first appear to be only a lipservice competition may readily develop into the real thing. The incumbent proposal writers no longer have the luxury to wait and see—they must write promptly and edit effectively to assure follow-up work, and to maintain control of the contract before competitors step in to arouse and agitate doubt. (With many service contracts the Federal Procurement Regulations (FPRs) or company policy require that recompetition should take place every so often, usually every second or third year. The recompetition renewal may merely be the legal observation of the requirement, or full-scale, combative competition.)

Briefly, the petrochemical study shows that an overwhelming majority of the clients, 67 percent in fact, perceived themselves to have contracted with a less than ideal incumbent, and admitted to having their confidence collectively shaken by would-be successors. Consequently, most of the incumbent proposal writers were forced to write follow-up proposals in a less than advantageous posture. These incumbent proposals, which traditionally have enjoyed a high ratio of success, were almost universally declined by clients in favor of total contractor replacement. Of the firms studied, 88 percent suffered this fate with the other 12 percent being scaled down in efforts.

In almost every case, incumbent proposal writers appeared to have written their proposals in a vacuum, with little more than intuition to solidify their positions. In synopsis, the defeated writers voiced their own shortcomings as to the points articulated and narrated in Figure 2.2. Here, there was a clear potential to remedy shortcomings, to cure client perceptions, and to hold onto the contract.

The Preproposal Process

In essence, the proposal writers whose firms were readily replaced had four key areas of mutual weaknesses. They failed to document to their clients that they did quality work; they failed to make a written personnel assessment; they overlooked or failed

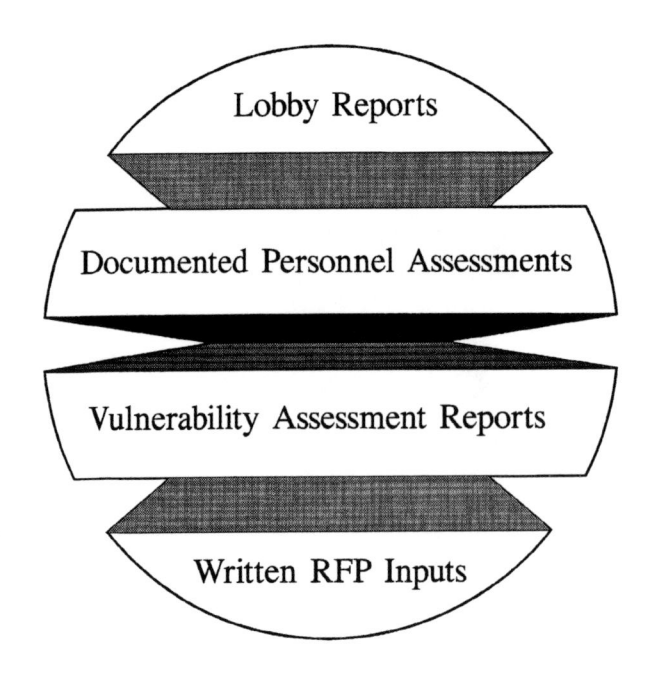

Figure 2.2
Preproposal Writing Activities

© 1996 by Holbrook & Kellogg, Inc.

to contribute significantly to the clients' request for proposal; and they neglected to conduct a documented *vulnerability assessment.* With few exceptions, it may be imprudent for an incumbent to omit any of the four steps, if that firm plans to succeed itself. In summary, these are the four facets of the preproposal process which require timely, efficient writing management.

Lobby Reports. Many of the losing proposal writers indicated that their loss was partially due to the client having lost sight of why the contractor won initially. Those criteria should be beefed up as a *lobby report* to the client prior to an RFP being drafted by the client. Original selection should be reinforced in terms of contractual achievements of an indisputable type, affirming incumbent strengths and overcoming what the client perceives as weaknesses. The time to win is *before* the RFP comes out, before competitors sense vulnerability.

It is imperative that incumbent writers remember that the competition is always hesitant about challenging successful performance, although they may go ahead to survey and probe for weaknesses. Winning incumbents will write up and disseminate their accomplishments; thus removing problems from the client's perspective before these fester. (See Table 2.1, which demonstrates just how polarized the competitors are from the incumbent.)

Table 2.1 Documented Incumbent Advantages *versus* Adversaries	
Incumbent	**Opponent**
Documented progress	No track record
Personnel assets	No staff
Revenue funded	Overhead funded
Experience on project	No project experience
Client good will	Unknown commodity

In short, when a contract must recompete, the client should be lobbied with written reports to assure that the client collectively recalls the prudence and judiciousness of the original selection. This documented lobbying activity should be done with the same determination and emphasis that won the initial contract.

Documented Personnel Assessments. Proposal writers usually win a new contract largely due to the personnel selected to staff that new effort. Depending on the number of years in a contract, those personnel have a tendency to migrate in search of advancement, client preference, or management prerogative. If key managers and key technical personnel have migrated out of positions initially agreed upon by the client when this contract was won, then the right people may no longer be in their right places.

A report on the imbalance (or balance) in personnel must be adjusted prior to the release of a new RFP, at which time many firms reported their adjustments were merely viewed as only superficial alterations.

Attrition of white- and blue-collar personnel must also be reported. What are root causes of people staying or leaving? If key personnel are leaving, this trend must be in print. Key managers who may leave due to weaknesses in their areas need to be propped up with worthy sub-tier managers, or in the worst case, replaced. Attrition may also be caused by competitors. Have any key people been hired away, and if so, by whom? Obviously, the proposal team should research and write this report in conjunction with the Personnel and Industrial Relations Departments.

If there is less than a year until an RFP could emerge, the contractor should mobilize its total resources, business allies, corporate friends, subcontractors, and suppliers for an overall personnel assessment. Without this kind of thorough report, the incumbent firm shortchanges itself.

Written RFP Inputs. All too many incumbents surveyed made no attempt to help the client write the new RFP, its specifications, or its requirements. In many climates, this writing effort may result in a *sole source solicitation* which is openly announced yet reserved for that one single unique contractor that had insight into the RFP. Along the Beltway in Washington, this process is sometimes known as "wiring the RFP," and equates to framing the RFP in terms to which only the incumbent can successfully respond. If the RFP cannot be prepared, the incumbent proposers can still inject their thought processes into the client's RFP as a psychological edge over the competitors to undermine their confidence with privileged information. The most successful route is to provide the client with recommended scopes of work and specifications.

© 1996 by Holbrook & Kellogg, Inc.

Incumbents' contracts and their initially successful proposals were likewise the objects of little attention, but represented major insights into client preferences. Here all former progress reports should be synopsized to demonstrate advancements. The contractors' archives should be closely monitored to assess how and why the initial contract and its proposal matured. Again, the incumbent proposers know the initial RFP better than anyone else. The criteria of a new RFP should be no surprise to successful incumbents who understand how their contracts are maturing and changing.

Vulnerability Assessment Reports. Universally, the proposal writers who were unsuccessful regretted not having gone over their vulnerable points. Overconfidence, excessive pride, and deafness to client criticism were cited as reasons for not having put together an assessment of where vulnerability existed. (Tom Peters, author of *In Search of Excellence,* refers to such unarticulated shortcomings as "corporate calcification.")

In the private sector, it is important to know who the naysayers are within the client's organization so that proposal writers can persuade them (again, early) that this incumbent's work is good, or at least acceptable. The costs associated with entertainment for this purpose should be accepted, contractors said, in the same light as judicious expenditures of business development funds that are merited for new business, certainly in advance of the RFP. These preproposal reports should be routed to the client early.

In the public sector of government contracts, the client agency's naysayers also need to be targeted for persuasion. However, while these people may be somewhat more resistant to having their opinions altered, they have very strict institutional and private codes of conduct that preclude direct, overt persuasion. A more subtle writing approach appears to be needed.

The agency's own civil servants are either right or wrong about their criticism of the contractor. If they are wrong, in the incumbent's opinion, then informal written presentations must be arranged to enlighten and revise client misconceptions. These briefings must be on an amicable basis, with no misunderstandings. If the civil servant in question does not respond, this process will have to be repeated to respective higher levels. This gentle pressure eliminates unfounded ideas, proposes written solutions to mutual problems, and removes points of contention—before potential competitors sense discomfort, disillusion, or outright hostility among some of the client personnel.

If, in the incumbent's opinion, the client agency is amicable (en toto) to contractor retention of the contract, success may not yet be assured. Of the firms surveyed, some 81 percent had unresolved audit agency and office problems at RFP time. Independent government agencies monitor federal contracts (along with the client agency's own auditors) and, in fact, have significant voices in whether to extend, rebid, or cancel contracts. Offices such as the General Services Administration, the Defense Contracts Administration, and various Inspector Generals are vested with such power.

It is imperative that the inquiries of these overseeing agencies receive in-depth written attention from an incumbent proposal perspective because such officials can inject enough doubt into the federal contract process to cause considerable contract vulnerability. The reports of these agencies are also double-threat documents because they chronicle accomplishments and shortcomings alike, are available to the general public, and are aired at congressional hearings. The adversary's proposal writers know how to reap these sources of vulnerability.

These agencies require another special preproposal writing approach aimed at decreasing contract vulnerability. First, proposers regret not having established a single responsible individual to respond in writing to all inquiries from audit agencies. This, they say, allows line managers to respond to ordinary work while giving a complete overview of what was requested in writing. Second, proposers must learn that the *modus operandi* of such agencies rests on assertions and rebuttals. That which is not rebutted is held against the contractor. That which the contractor overlooks (or omits, for any reason) is likewise held to be true, and therefore to the detriment of the proposer.

This dangerous system of argumentation needs to be met on a one-for-one basis, documented by the proposal team with strict attention to giving each agency a straightforward, positive answer to each single inquiry. Blanket answers were shown not to suffice, and drew the ire of agencies in some cases. Audit reports from the oversight groups must also be scanned methodically by proposal writers to uncover covert criticism and rebut it. Overt criticism was felt to have been better counterargued promptly, preferably with enough technical data to erode unfavorable assertions.

© 1996 by Holbrook & Kellogg, Inc.

A final point—oversight agencies are today not limited to the government. More and more private firms have auditors for their subcontractors' cost, schedule, and performance. It is again imperative that these privately employed overseers receive the same written, documented attention paid to government audit agencies.

Incumbent proposal writers can and should begin proposal activities far in advance of the RFP, with special emphasis on the important areas of lobbying reports, personnel assessments, RFP inputs, and the vulnerability assessment study. Again, incumbent proposal writers are in the optimum position to guarantee the successful continuation of a contract by eliminating perceived unsatisfactory performance prior to competitors exploiting such data in their own proposals.

2.6 THE MANAGEMENT-MARKETING PLAN

No proposal can be successful without marketing; no marketing effort can be successful without a marketing plan. A succinct plan covers three to five years. Three-year contracts in progress thus fall into perspective, as do contracts one to two years in the future. The true five-year plan encapsulates the past, present, and future. The five-year plan is an overall plan for setting goals; for example, where do we stand in the industry, and where do we need to be? The president and vice presidents of corporations typically outline these goals.

The three-year plan is a quite immediate, specific, practical plan during which the company assigns people to new projects, writes feasibility studies and white papers, and persuades clients. Department or division managers usually assemble this kind of plan.

One-year plans actually make up an annual blueprint of who will be dedicated to what projects, problems, and proposals. Line supervisors and field managers put together these fine details. Progress must be made in each of the three types of plan, or priorities should be redirected toward more promising projects.

The *quarterly plan* is the most specific, and outlines the following: aiming for current proposals; traveling; making arrangements with subcontractors and vendors; and above all, concentrating forces where marketing shows the most proposal successes will occur. The quarterly plan must be reviewed and approved by each management level.

There is a certain amount of unexpected RFP activity that comes from clients whose actions cannot be well predicted. Consequently, each marketing plan should contain and categorize RFPs that could "pop up" with little or no warning. Every quarterly part of the plan should have a cushion factor in case one of the so-called "targets of opportunity" occur, rushing the company into the proposal mode hurriedly. Thus, any quarterly marketing plan will show real and projected targets, allowing enough manpower and coverage to show a best-case scenario; if the RFPs are indeed released simultaneously, the firm will have to respond accordingly to a worst case set of events.

GANTT charts lend themselves well to such planning activities, and software like Harvard Total Project Manager will show planners where overlaps and crises are likely to develop. All marketing plans should be automated, showing the full range of probable events. Again, the quarterly plan will yield the most immediate management results, showing expenses to be incurred, crises, and slack time—that all important cushion when the firm reorganizes its resources for the next effort.

2.7 CONTRACT MARKETING ARCHIVES

A firm without archives is blind not only to the competition but to its own capabilities. The archives should trace the firm's capabilities and qualifications for the last five to 10 years, including annual reports, realignment or reorganization reports, contracts on which the firm bid, and win-loss information. It is essential to have documented evidence on hand as to why a victory or defeat occurred.

Parallel files should reflect the progress of each major competitor, and should consist of any competitors' proposals obtained from new employees, allies, or the Freedom of Information Act.

White papers and data on research and development (R&D) should also be kept current within either the marketing archives or the company library, but employees should not be allowed to take these documents home. A succinct effort to assess the win-loss ratio must be accessible to the proposal and marketing teams, so that corrective actions can be taken.

2.7.1 Proposals in the Aftermath: Postmortems[**]

Many technical firms can benefit from a proposal postmortem function. Aerospace firms regularly perform this kind of assessment (and the term "postmortem" comes from that industry), but the technique can be well used outside aerospace. For example, a new, small firm is more vulnerable to traumatic losses than larger firms, yet it does not have the resources to conduct postmortems. It is possible to scale down formal postmortems to brief but succinct reports in order to obtain a complete but swift compilation of data.

Postmortem sessions focus on the official communications between the firm and its client. They emphasize the successful portions of the writing, with an eye toward improving the unsuccessful. They address strong points, weak points, and cost in a formal atmosphere free from so-called postaward euphoria.

Keeping the company's proposal archives updated via the postmortem process improves training resources for proposals and thus has a beneficial effect on teams engaged in proposal writing. Costs are also more meaningful in this cool, detached view of what has been achieved at what price.

The postmortem analysis is best conducted by a separate team. In a traditional proposal effort, there are only three specified functions:

- **White Team:** Develops the technical approach and writes the proposal, using a substantial amount of technical publications assistance.

- **Red Team:** Redlines and critiques the proposal efforts in draft, using additional technical publications aid.

- **Blue Team:** Assesses the opposition and provides ideas to defeat the opponents' strengths and exploit their weaknesses.

But an important team is missing—the Gray Team. Its members should consist of at least one representative from all three teams, but there should also be one or more detached, uninvolved persons to help ensure an objective verdict.

[**] This section is based on an article from *Technical Communication*, 2nd Quarter 1986.

The manager of a separate proposal team makes a good Gray Team manager. The manager must be as familiar with the RFP's intricacies as the White Team is. The Gray Team manager must assemble all proposal data and conduct the review sessions. (If the Red Team is there to add red ink, the Gray Team is there to remove gray areas.)

A good way to collect the data is to put them into a Gray Team summary sheet. (Figure 2.3 is a good example of a thorough de-brief sheet from an aerospace Gray Team.) The Gray function pores over all available information, formally and informally, including:

- Source selection board documents;

- Client reports, letters, and debriefings;

- Records of client phone inquiries;

- Analysis of the winner's qualifications;

- Disposition of Red Team remarks; and

- Other data (Figure 2.4 illustrates typical contents of proposal archives.)

GRAY TEAM SUMMARY SHEET

Proposal Name: _____ Project Number: _____

       Won ☐   Lost ☐

Proposal Manager: _____ Total Proposal Cost: $ _____

Was Proposal Rated by Client? _____
                                     (Example: 2 of 5, 1 of 3, 3 of 3, etc.)

Contract Awarded To: _____ Contract Value: $ _____

Was PRICE the acknowledged winning factor?   Yes ☐   No ☐
Was Technical Superiority acknowledged winning factor? Yes ☐   No ☐

DEBRIEF SUMMARY: Describe Source Selection Board's remarks, and attach a copy of that letter to this sheet. Highlight reasons articulated and not articulated by SSB. Include Gray Team findings. Use additional sheets if necessary.

Business Development information accurate? Comment:	Red Team Remarks used for revisions?

Gray Team Members: Gray Team Manager:
                                      Name/Title: _____
                                      Date: _____

Page ____ of ____

SENSITIVE COMPANY INFORMATION, LIMITED ACCESS

Figure 2.3
Proposal Postmortem Summary Sheet

Proposal Archival Data Sources

- Formal Award and Debriefings
- Client's Informal Comments
- Orals and Inquiries
- Other Feedback
- Request-for-Proposal (RFP)
- Subcontractors' Information
- Published Data
- Lobby Data
- Business Development Data

→ PROPOSAL ARCHIVAL DATA → Gray Team Findings & Reports

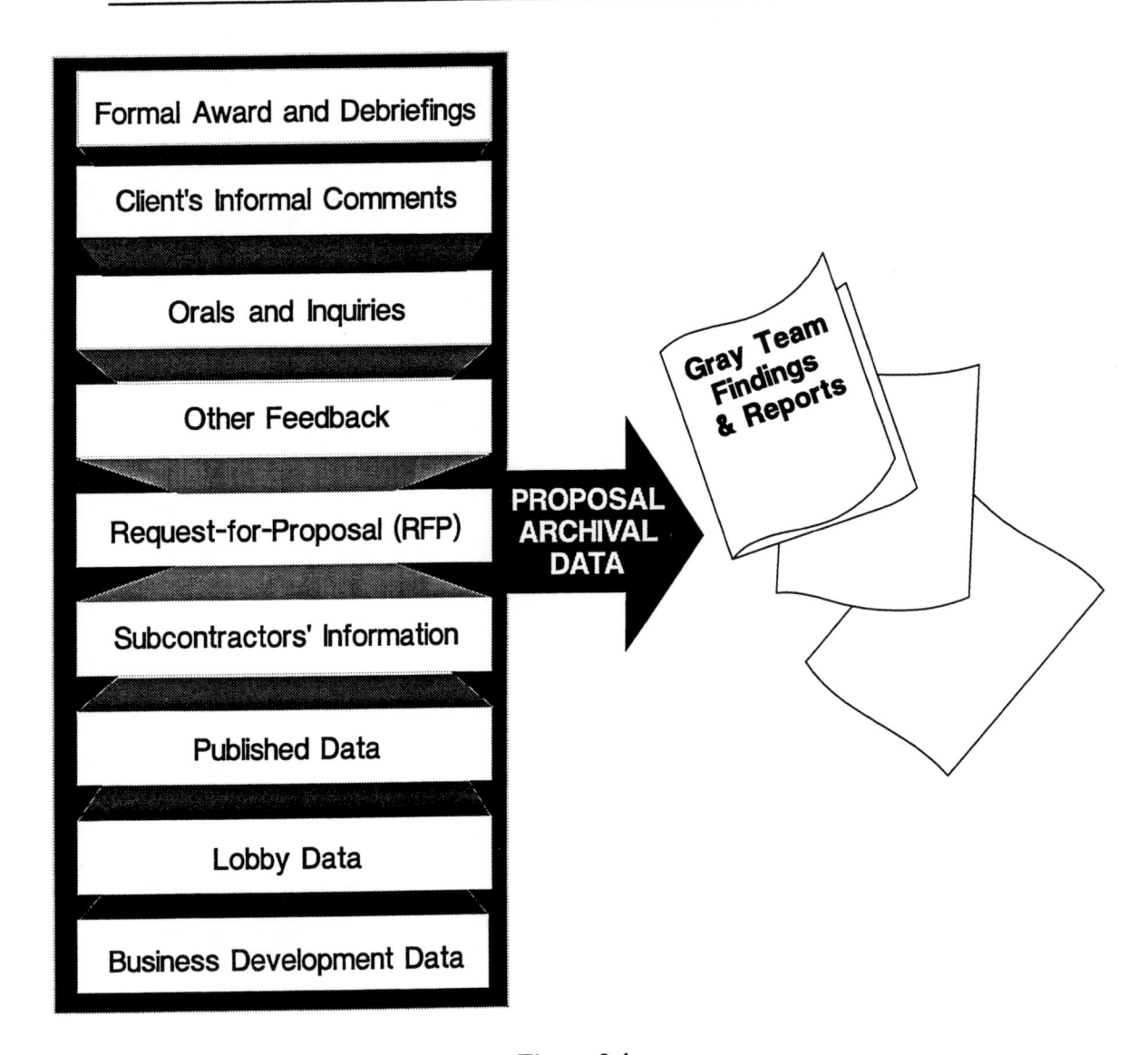

Figure 2.4
Archival Proposal Data

A Gray Team manager should be skilled in reading bureaucratic jargon, and should recognize that official source selection board remarks are written by committees who are proud of saying nothing, but saying it well. Reading between the lines can sometimes provide clues to losses, whereas winners usually are told simply "best price," or "technical superiority." It has been said that board members are looking for losers as they evaluate, and those proposals that escape looking like losers will creep into the winner's circle. Another view is that a clear winner emerges early, and it is then necessary for the board to develop reasons for the other proposals to lose.

In either case, a great deal of personal judgment goes into the board's verdicts, despite the appearance of a quasi-mechanical evaluation process.

If the Gray Team manager detects serious flaws in the board's reasoning and feels that the company was unfairly dealt with, the manager may recommend a formal protest, with the possibility of overturning a selection board's faulty judgment. There are no sour grapes in this activity; it is a normal part of the procurement cycle, and an appropriate Gray Team function.

The Gray Team eventually produces a brief report for the archives, assigning reasons for each victory or defeat. At this point, costs can also be critiqued from engineering and publications perspectives. Typical findings reveal that time and resources handicapped the effort, that some sections were well prepared—but resulted in skimping elsewhere. Often, management will overrule technical personnel in a section, or *vice versa;* these conflicting views need documentation.
Sometimes agreed-upon emphasis faltered; what caused this alteration for the worse? Another key flaw is inaccurate, overly optimistic business development data. (Many firms are guilty of this flaw, as well as of coming up short on inside information.) One more typical flaw is the willful disregard of Red Team critiques because of pride of authorship. Text that looks awkward to the Gray Team should also be identified to coach its writers. Better passages should be praised.

The Gray Team articulates the compliance and noncompliance of the proposal with the request, and assesses whether technical know-how was adequately shown. With this information available, winning performance can be repeated and losing efforts can be improved. Costs will be scrutinized as well. In some cases, firms found they had invested too much time, by 50 percent, on a proposal and thus drained their resources.

The Gray Team is in the best position to see how well the Red, White, and Blue Teams did, and what they got for the company's money. Summaries of time sheets, consultant fees, milestone dates, and management directives will be needed to ensure an accurate picture. For example, what was spent on the technical approach section? That is typically where the emphasis (and costs) should be heavy.

Like projects, proposals should be on time and on budget. Time and cost overruns usually spell faulty task management. If time and resources are lavished on small issues or an unproductive goal, the Gray Team should discover this disproportion.

© 1996 by Holbrook & Kellogg, Inc.

Other shortcomings include research into materials that were unusable in the proposal, or topics so esoteric that even in-house experts are baffled. Similarly, it is poor management to say, "We need 500 pages of technical data, so 50 people writing 10 pages each will do the job." Or to say, "The Quality Department is available. Let's get them to write it for us."

The result of either situation is usually a mismatch of work to people. If managers pay the same price for 500 pages of rambling, imprecise verbiage as they would for 250 pages of succinct text and art, then they deceive themselves. Extravagant overtime and consultants may also enlarge the proposal's price, with a limited payback value in quality.

One final task of the Gray Team should be a last evaluation of the available preproposal business development information to see if a no-bid would have been a better economic choice. In the heat of the proposal writing phase, the wisdom of a no-bid decision is often overlooked.

2.8 SECURE PROPOSAL OPERATIONS

The ultimate goal of a marketing department and a proposal development group is to have a current, vital data base of information germane to winning new contracts. This is a valuable commodity, and one to be guarded jealously. The marketing and proposal files should have adequate security to preclude unauthorized access, losses, or unrestricted copying. There is a price tag on every piece of marketing information and every proposal. The firm made investments in each document, and should regard those papers as part of the company's most valuable assets. A badged access policy is advisable, enabling only managers to sign materials in and out. Access to copies should be limited to only the proposal writing teams.

When an employee leaves for any reason, his or her personal proposal papers should be made a part of the termination or retirement process so that valuable data do not inadvertently fall into the wrong hands. (See Figure 2.5 which shows all the elements of good security precautions for proposals.)

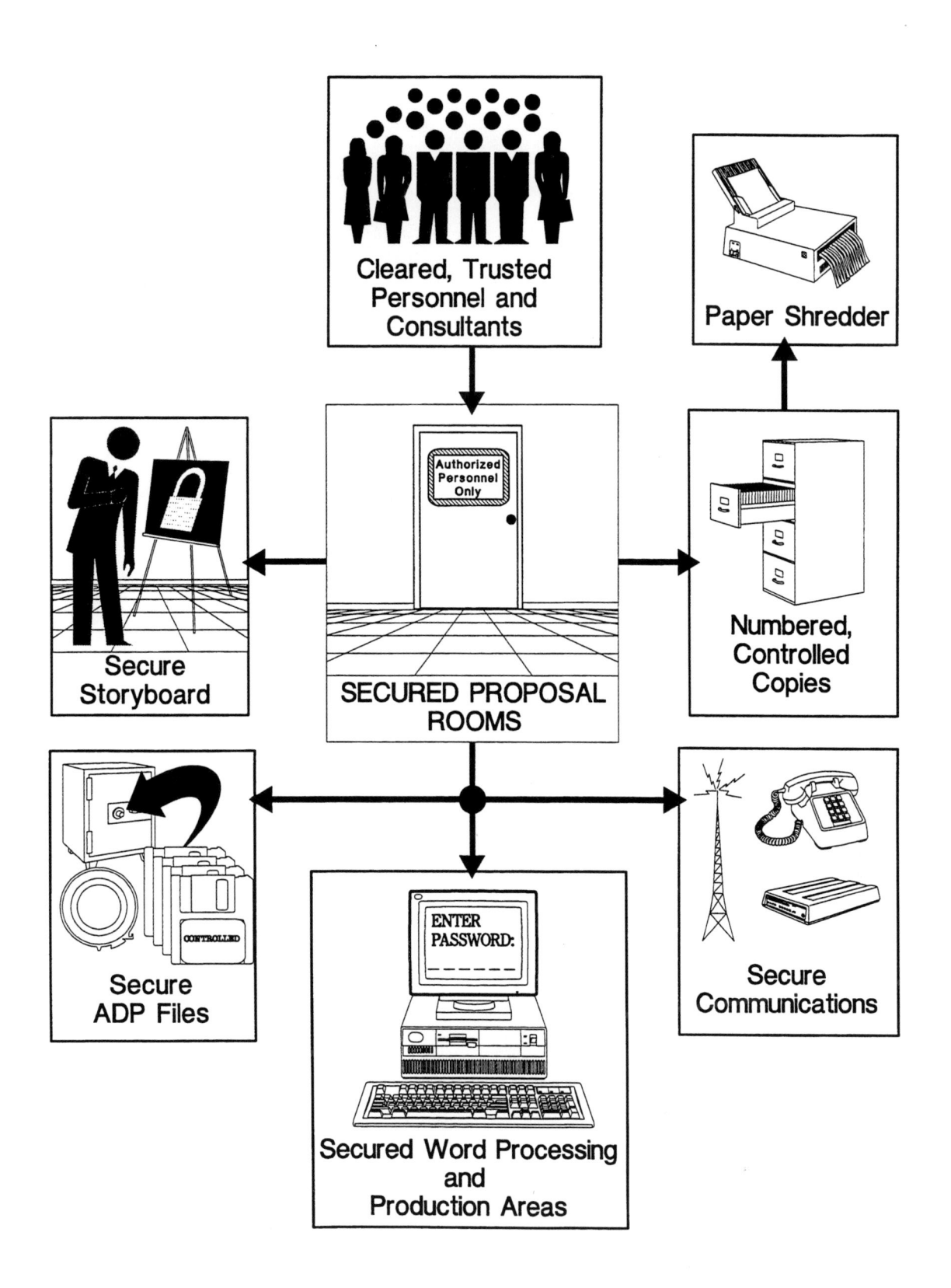

Figure 2.5
Security Flow for Proposal Development

2.8.1 Security in Proposal Operations

We should make a final point on proposal preparation security. Because of the large dollar investments made in proposals, the documents are valuable, and consequently they are targets for theft by competitors. The average aerospace proposal in 1994 often costs over one million dollars for the three to six months of work by the managers, engineers, technicians, drafters, editors, and others who realized the proposal. That energy and expense reflect what must be a secure company investment, without leaks, gaps, loaned or borrowed copies, or excessive numbers of file copies. However, whereas control of final copies comes at the end of the proposal process, security is a responsibility throughout the entire process.

Primary Security Features

A good way to handle proposal security is to proceed along the provisions of Army Regulation 380-380 on computer security, which imposes the following beneficial limitations:

- A formal list of persons in the company authorized to participate in the proposal—access to any proposal area will be dictated by the list only;

- Badged access if actual government confidential, secret, or top secret material is present in any way;

- Persons should have a strictly specific reason for requesting information from others;

- Locked conference rooms and office areas behind which unauthorized persons may not enter;

- Daily sign-in and sign-out logs, for all persons;

- Locked rooms for storyboards, word processing, and graphics;

- Locked files and desks;

- Spot checks of desks, files, briefcases, trash, and outgoing documents, and the proposal manager is made aware of shortcomings and violations;

- Special, centrally located wastebaskets to collect sensitive waste papers;

- Shredders in the area, or daily removal and destruction of scraps, redlines, and other surplus pages;

- Visitor control (visitors must be approved by a company official to limit superfluous outside contacts from vendors, subcontractors, or consultants); and

- Control, disposal, and eventual shredding of Red Team and Blue Team draft copies and intermediate issues.

Computer Security Features

Perhaps more important than limiting access to paper is limiting access to computer generated designs, simulations, trial programs, data bases, and other computer reports. In the army regulation mentioned earlier, areas for control of data are as follows:

- Individuals are screened for active company security clearances or government clearances;

- Computer room and terminals are protected by combination or by locking doors;

- Locks and combinations are known only to top managers;

- Passwords to computer data are assigned by the proposal manager and volume managers only;

- Algorithms, codes, and test routines are controlled;

- Simulations and models are controlled;

- Printouts are limited to specified employees;

- No tie-ins to remote terminals are allowed;

- Secure telecommunication links;

- Printouts are restricted to other control requirements, and shredded or destroyed when they are no longer useful; and

- The company's security department periodically inspects records and transaction files to ensure freedom from "hackers" who are inquisitive or employed by competitors.

As a general rule, proposal security will become lax unless the proposal manager insists on it from the first day. Volume and task managers must demand controls over paper products such as drafts and printouts. Another aspect that requires control is conversation, which often inadvertently betrays approaches, design, concepts, personnel names, prices, and other key strategic items. Beware of vendors, subcontractors, and consultants who are inquisitive or volunteer to help, uninvited. Idle talk can pass along enough information to help a competitor grasp strengths and weaknesses, and then exploit them.

Understanding the Competition

Let us consider the other side of the situation. What should we do to penetrate the enemy camp? How should we get into the private domain of our competition? Some of the more creative attempts to breach the security of others include the following unethical activities:

- Job offers to key managers and engineers three to six months ahead of the RFP;

- False information about your poor performance planted with the client to seek cooperation from one or more of the client's staff in leaking such data to the industry at large;

- Friendly people at local bars;

- Bugged telephones, or listening devices in hidden places;

- Phony trash collectors, janitors, and employees from "another branch";

- Sound gathering devices aimed at key offices; and

- Bribes to print shop personnel to run one or two extra copies of drafts and finals. (By the way, who supervises your proposal production area when copies are run? Who controls the originals and *disks*?)

An intelligent proposal manager warns the team that there are unscrupulous people out there. The penalty for carelessness is losing new contracts. Employees must refrain from careless talk at lunch, after work, and on the phone to prevent leaking important marketing data. In short, here are the characteristics of security conscious employees:

- They are loyal to the company;

- They have made few if any security errors or breaches and they do not complain about security;

- They are happy, satisfied, dedicated, and uncompromising;

- They dislike the competition;

- They are not loners, or prima donnas given to pouting or despair when a decision goes for or against one of them;

- They work a minimum of eight hours a day, and rarely leave the proposal area; and

- They have incentives to win, like a victory bonus.

An Example — The Walk Around

One method the proposal manager can use to manage security is to walk around. Let's suppose an ordinary trip through the team yields the following kinds of problems on a daily basis.

The problem: The engineers have amassed a pile of papers that aren't shredded because the shredder is broken down, again. Papers are going into the ordinary trash for this reason. There is the potential for a major compromise and leak.

The remedy: Hire a bonded shredding firm to pick up and destroy the data, under your supervision. Keep your own shredder maintained and ready to go. Assign someone to shred every day, picking up data from the team members, one-by-one.

The problem: Technicians are chatting openly on the phone about what they're working on with their friends.
The remedy: Post a memo every Friday repeating your goals of productivity, and ending with "Keep your phone calls limited to non-proposal matters. Do not discuss our work with others."

The problem: The receptionist or secretary is at lunch and entry is unchecked.
The remedy: Make a roster of people who double as "guards" or backups for the normal staff so that entry and exit points are always controlled. Audit the sign-in log for the initials of the people who vouched for their identity.

The problem: A tired manager wants to take home the technical proposal and read it again over the weekend.
The remedy: Many firms flatly forbid removing data; others insist on hand receipts from trusted career employees only. The least level of control is numbered copies, plainly marked and briefed as *Company Confidential/Not To Be Reproduced.*

Generally, the tighter the security restrictions, the better. Only the proposal manager and designated volume managers should be trusted to audit the secure work of others; the ordinary proposal contributor will be so involved in doing the writing that control of the paper and media will be only a passing matter to them.

In closing, the proposal manager has to act quickly to resolve any potential security leaks. Persons who regularly fail to adhere to security policy cannot be tolerated in the proposal preparation loop.

2.9 FEASIBILITY STUDIES AS MARKETING TOOLS

In the rush to make frequent marketing calls and write proposals, one traditional method for acquiring new business often is overlooked: the *feasibility study*. Such a study exhibits characteristics of both a detailed proposal and a miniproject.

The major marketing and business development asset of a feasibility study is that it can demonstrate that a client's project is practical. Because such proof often is of paramount importance to a client, he usually is willing to pay for such an effort. In proposals, most contractors merely state that they will follow the client's scope of work and specifications. However, a firm that offers to prepare a feasibility study tells the client that a detailed scenario of the proposed project will be developed and that simulation will be a good investment of the client's money because it will show what the client can expect, both technically and in cost.

Targeting the Study

Persuading the potential client to embark on a feasibility study is not difficult, but timing is important. If the potential client is deluged with other proposals, the client will have to sort through many would-be contractors, many of whom will promise lavish work for small payment. Therefore, the firm offering to perform the feasibility study must selectively target those clients whose projects have a history of fulfillment, and make the offer to conduct the feasibility before an RFP announces a project to the engineering community.

By intercepting the RFP, the feasibility study enhances the client's knowledge of the work to be performed, and prescreens potential contractors, thus easing the client's responsibilities. The firm that offers the feasibility study must scrutinize precisely who its competitors are, and structure the study to reflect the firm's own strengths, its competitors' weaknesses, and the trade-off. At this point, the feasibility study represents the firm as the owner's executive engineer, recommending courses of action to achieve the project's goals.

The goal is, of course, a final, full-scale contract. This often can be achieved after a successful feasibility study is completed. In fact, acquiring such a contract is almost a certainty because the unknowns that can plague a new project have been analyzed thoroughly. (A note of caution: some studies are so detailed that the owner no longer needs the engineer, an error that is easy to make. It is imperative that the engineer show enough detail to prove thorough understanding, but not enough to make himself dispensable.) (See Figure 2.6, a succinct view of how multiple approaches have to boil down to a single approach.)

© 1996 by Holbrook & Kellogg, Inc.

The steps in this advance from a proposal to a contract are the following:

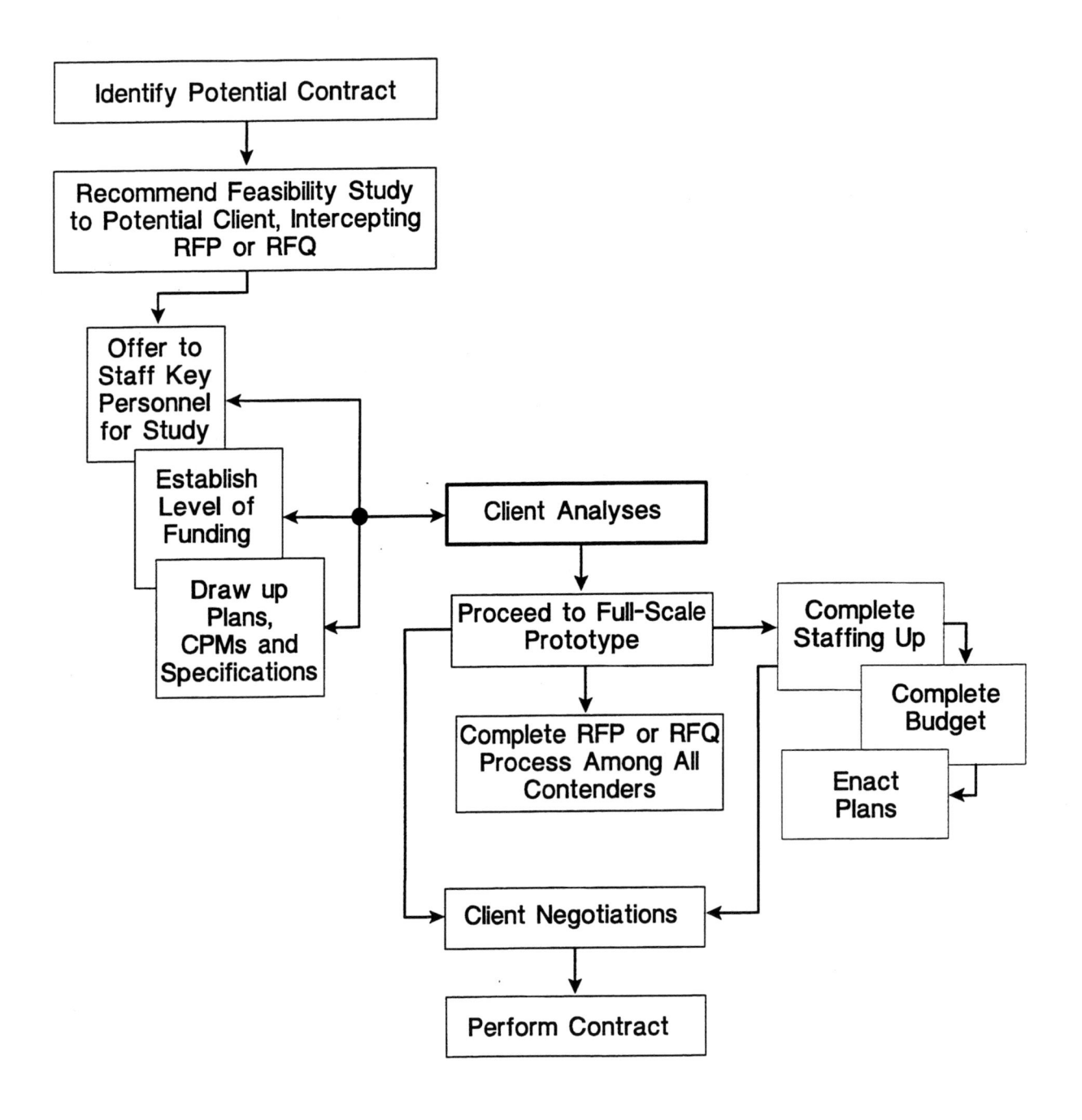

Figure 2.6
Initiation of a Feasibility Study

© 1996 by Holbrook & Kellogg, Inc.

- *Proposal:* a brief look at a scope of work, specifications, job site, and inherent problems.

- *Feasibility study:* intense, elongated overview with ample planning and analysis, with enough time for delegation of responsibilities.

- *Contract:* full-scale development, either by proposals only or with the benefit of a feasibility study.

For the firm that conducts feasibility studies on behalf of its clients, the positive factors are numerous. The offer to perform a feasibility study may intercept the client's issuance of a solicitation for proposals and, if accepted, may focus and limit consideration for the work only to the firm that prepared the feasibility study. Also, an immediate start on the client's problems can be made because the scale is limited, and there is no extensive commitment of client funds. The study will be less theoretical and less academic than any proposal—a proposal is, after all, only a set of elaborate promises. In addition, specifications, critical paths, and plans can be developed that will lead to a full-scale contract.

The Possible Disadvantages

However, we must be aware of potential disadvantages, which can be ameliorated by good management techniques. There may not be enough regular staff employees to detail precisely what would be expected of a full-scale project effort. The work produces less revenue than a full-scale project, and key people will be committed to the feasibility study instead of to other projects.

Fortunately, many of these potential disadvantages can be used in a positive manner. A large percentage of management time will guarantee the client that it is receiving optimum management attention to its study. The need for adding more staff may be seen by the client as an opportunity to ask the technical firm to enhance its efforts, going logically toward full-scale development. Low revenue need not be a problem because low costs can be equated with low risks. Lastly, the commitment of key people can be thought of as an investment in reimbursable talent, which is available and already acclimated to the project. This is important if the client commits to full-scale development.

From a cost-estimating perspective, the engineering firm has a very firm grasp on what budgets should be. From a skeletal staff, the company can extrapolate what full-scale staffing will be, then add the company's normal overhead, general and administrative costs, fees, and profits to arrive at a realistic bottom line for staffing. Support, office space, and ancillary items then can be estimated based on staff size. The technical analyses developed during the feasibility study can be used to estimate the forthcoming project, thus arriving at overall personnel-hours.

The cost of the feasibility study itself should be based on the firm's normal rates, plus overhead and general and administrative costs. There should be no effort to offer a client a bargain-basement price because that kind of strategy only calls into question the credibility of the firm and lessens the reputation of the profession. Any client who needs a feasibility study can pay for one at the normal price.

The client should get a quality product for its money. No effort that would normally be expended for a good client should be withheld from the company that is paying for a feasibility study, and its vouchers and invoices should reflect attention to detail, as well as cost consciousness for any item that might be considered superfluous to its immediate needs.

It should be noted that this method of acquiring new business has been polished to near perfection by 8(a) small businesses, many of whom can gain access to Federal agencies with a fraction of the red tape that larger firms encounter. Partnering with an 8(a) firm is an excellent way to share the risks and successes of that avenue of gaining new business.

From the firm's point of view, a feasibility study is an economic way to establish expertise, avoid the pitfalls and expenses of competitive proposals, and perhaps create a new, if limited, reimbursable account.

CHAPTER 3

Sizing Up the Competition

The competition is lethal. At one time or another, all marketers experience the disappointment of losing a "sure thing" to a competitor whose last minute efforts squeaked by to win. Usually, the competitor's victory is the product of superior marketing intelligence, but price and political connections have also captured other "sure things." No successful marketing staff can expect to prosper without significant insight into the competition.

There are many ways to get information so that we can evaluate other firms. For example, car manufacturers and computer mainframe manufacturers are numerous, but not too numerous for most interested people to list and make a few value judgments. Someone who is determined to buy a new car or computer will make an effort to learn about the finer points of performance, prices, reputation, services, and repair, as well as other measurable characteristics. Any firm that sells goods and services has these avenues open to its customers, and to others who are not customers but who are interested in learning about existing paths of information.

3.1 DATA BASE—PUBLIC AND PRIVATE

Competitor data bases are files of information taken from what is available from firms through all sources. To beat the competition, we must know these sources. The opponents cannot remain an intangible, shadowy commodity if we understand their shortcomings, strengths, and how they win.

© 1996 by Holbrook & Kellogg, Inc.

Annual reports, capabilities and qualification statements, Dun and Bradstreet ratings, general credit ratings, stockbrokers' profiles, and data from past and present employees, vendors, consultants, and subcontractors all contribute valuable information. Who will give us accurate data? Business allies are one source; roving consultants are another. Freedom of Information Act requests are another way to get data, but we are required to state why we want the information. We are looking for a pattern of behavior for the competitor, so that we can predict the competitor's future behavior.

The Freedom of Information Act

The Freedom of Information Act (FOIA), or the so-called Sunshine Law, is often the key to obtaining accurate information about an opportunity. The agency funding the contract that interests your firm should be screened to learn what materials exist in the form of former and existing contracts, change orders, amendments, and prior RFPs that will illuminate the work. Many firms will not seriously consider an opportunity until and unless they have on hand all the data available through a FOIA request.

Each firm that expects to score major successes in new work, as well as to defend their own contracts, must be competent and aggressive in getting results from their FOIA requests. If your contracts or legal staff are not writing these requests for your firms frequently and routinely, they should be. Without the flow of FOIA data, you are in a genuine information black-out. As you will see in Section 3.1.2, Audit Agencies, you should also FOIA audit agency reports of contracts that are in dispute or experiencing fraud, waste, or abuse—real or imagined. In brief, firms that win often, FOIA often.

Public Information

A blend of public and private data—for example, a small library—is essential for assessing a firm's position in its industry. Here are some of the items and indicators that provide public scrutiny:

- *Annual reports:* where do the competitors rank themselves and what do they brag about? What is their premier line, product, or project? Did they make any money last year? How many people do they employ? What goals do they have, and what did they assume to project such goals? Do we have their latest annual report?

- *Capabilities and qualifications statements:* what do they claim is their expertise? What made them well known in their segment of their industry? What job skills, special engineering skills, and personnel do they have? What are their current sales figures and their backlog? Do we have their latest capabilities statement?

Let us consider significant documents that are difficult to obtain, such as Dun and Bradstreet ratings, general ratings for credit and reputation, and private assessments. The annual reports and capabilities booklets are almost always glossy, optimistic accounts of glowing achievements—they omit or disguise bad news like unsuccessful projects, forced reorganization, slow payments, multiple lawsuits pending, and the character of top people.

Dun and Bradstreet and stockbrokers' subscription files give a good idea of the financial status of a firm. Are your competitors subject to slow payments to subcontractors and vendors (which signals poor cash flow management)? Do they deal with frequent workman's compensation suits (poor safety in the workplace)? Do they reorganize frequently or receive Federal court orders (poor management)? Have they had contracts terminated; stop work orders; safety violations from OSHA; or fines, penalties, and judgments (which indicate a general state of malaise)?

A firm that fits these descriptions is beatable. On the other hand, a firm with excellent project performance, AAA credit, no litigation of any significance, and solid management track records is a tough competitor. This firm will have the kind of *similar experience* that will inspire source selection officials to choose it as the winner. It will have the performance attributes on which contracting officers can gamble their careers. These firms are good performers—the kind that each company aspires to be, and the kind that clients love to recommend. Good *performance* is the heart of such a successful reputation.

Private Information

Private assessments offer the most reliable insight into the competition. Marketing strategists have to know the answers to these questions:

- What work have the competitors done?

- Who is leading their proposal efforts?

- Who are their chief engineers?

- Who has worked with their people?

The marketing network should reach for all data on the competitors' experience and client contacts. We must take preemptive steps to position ourselves to win. We should assume that our competitors use the following tactics:

- They want to predict what our proposal will say;

- They will try to hire away our most competent employees;

- They will try to penetrate our proposal team;

- They may buy borrowed intelligence;

- They will interview as many of our past and present employees as possible;

- They make their intentions well-known;

- They may employ unethical tactics to exploit or negate firms that they consider serious competitors; and

- "Fronts" may buy actual goods or services from us, then disassemble the product or rigorously analyze the service.

In essence, the marketplace contains many ruthless competitors, many of whom win largely due to their intelligence gathering operations. Sizing up the competition is much like a reconnaissance aircraft flying over strange territory in an attempt to find an ideal beachhead. Wrestling for new business is no less sincere than campaigns in wartime. The trick to knowing what to expect of the opponents is not to look at them from one perspective or vantage point, but from many.

© 1996 by Holbrook & Kellogg, Inc.

When we see the competitors emerging as real people with certain skills and some flaws, we are getting close to knowing what to stress in our proposals. We should seek the opinions of many people working with the competition, not just one person. Also, we should assign at least two employees to analyze each competitor, so we can discuss and agree on what they saw and heard.

When public and private data are merged, a true picture of what the competition is like will emerge. Thus, the opponent's promised performance can be dealt with later in the proposal. We must keep a file on every major competitor, and update that file regularly.

3.1.1 Congress and Congressional Agencies

We have a friend in court, after all. Congress and the government oversight agencies provide us with much of what we need to know to market effectively and write winning proposals. A significant part of each annual marketing plan should be devoted to agency-available information. Better yet, we should use state and regional representatives in Congress, who wish to chaperon and safeguard local interests.

Lobbying—A Regulated Industry

Each senator and representative has access to the entire Federal system, and the marketing staff has to align the firm's interests to those of selected officials. The best way to achieve these results is to do the following:

- Hire an accredited, registered, professional lobbying firm;

- Hire an individual lobbyist who has an ethical reputation; and

- Team with firms that have a quality lobbyists.

To be sure that your lobbyists are getting the right data, we must make special efforts to double check all their work. To do this, we assign two to three key managers the task of regularly visiting the local and Washington, D.C. offices of our senators and representatives. Also, we should compare our notes and work assignments to our lobbyists' assignments. The lobbyists' time sheets should reflect fair charges for work. We should start an audit trail, so our accountant can track where our investment went, and for what. The Department of Justice, in Ill-Wind cases, cited numerous lobbying abuses to avoid. Contractors must ensure that important programs get appropriate attention in Congress, using ethical, lawful persuasion.

Tracking

Additionally, the firm itself must be responsible for tracking procurements through Congress, the Pentagon, and other Federal agencies. Responsible, politically skilled managers should track congressional calendars and hearings, the *Congressional Record, Congressional Digest,* the budget of the United States (and each subbudget), and maintain a record of where budget dollars are going. (Some services in the Washington, D.C.-Baltimore area now offer such data in an automated manner, plus bill tracking through both Houses of Congress.)

As budgets solidify, the firm must know where the government's general interests lie, and how much funding will flow into each interest. (If your primary clients are large contractors, you will need to scale down your marketing to their specific targets.) Once the overall annual budget picture is seen and comprehended, we will know which military and federal agencies have the financial means to procure goods and services. (At some consulting firms, this step is known as "feeling the purse," a reference to sizing up what the client has to spend.) An approved and funded program means it is a good prospect for work in succeeding fiscal years.

Counterparts

At this point, conscientious contractors may schedule new work two to three years ahead of actual procurement, well before the Federal agencies formulate requests for proposals, scopes of work, and other government solicitations. At this point, the search for *counterparts* begins. Counterparts are those officials who are responsible for the agency's budget. They decide where and how to spend the money in the budget. Counterparts are the GS-14 (and higher) civil servants, and the so-called supergrades of civil service who are politically appointed.

The counterpart plays a very significant role in large procurements. Contractors first emphasize the quality of their goods and services to noncongressional individuals—that is, people who have specific, nongeographical motivations linked to the performance of new work. Counterpart personnel size up contractors to design the next ICBM, launch the space shuttle, or take Star Wars into the next phase.

Counterparts are the pioneers or pathfinders of agencies, and they forecast who is to be the *front-runner*. That is to say, if you were a career civil servant, devoted to a safe, secure future for yourself and your agency, on which contractor would you care to wager your reputation, chances for promotion, and hard won success? Who do you consider the special firm that will neither let you down nor embarrass you, the reputation and integrity of which you respect?

Congress allocates funds to projects and people who have concrete concerns. Each agency has a management structure which roughly corresponds to that of the counterpart. People are named responsible for allocating the funds, and have to report to Congress on how they will distribute that funding into engineering designs, studies, operations, maintenance, and other typical contractual relationships. It is at this level that the firms should send key managers to the counterparts to see what opportunities will arise in one to three years, and gauge what concerns, ideas, and concepts the counterparts have about how to proceed with the work.

A constant transfusion of ideas and knowledge flows between these knowledgeable contractors and their Federal counterparts. It is through these counterparts that contractors first identify, locate, and influence technology. Without this high-level step, contractors are dependent on the good will of more thorough firms to succeed. It is essential to secure congressional patronage at the earliest moment, and to cultivate that patronage continuously and emphatically.

3.1.2 Audit Agencies

For Federal acquisitions, many contracts go sour for each one that succeeds. In fact, one nonmilitary Federal agency in Maryland always has 10,000 blank contract termination forms on hand. The government client agencies assess success and failure through formal and informal investigations for Congress. We must take advantage of these important oversight agencies: the General Accounting Office (GAO), the Inspector General (IG), and Defense Contract Audit Agency (DCAA). We need to know how to use the reports of these agencies—the opposition already does.

It is the charter of the audit agencies to report on how well each contractor performs. On a regular basis, contractors are cancelled for the following mishaps:

- Accidents, near accidents, deadly situations, and actual deaths;

- Fire, faults, and chronic errors;

- Cost overruns;

- Slips in performance, poor achievements;

- Sloppy accounting of milestones; or

- Fraud, waste, and abuse.

The audit agency is valuable to the intelligent contractor because of its reporting function. Is the competitor performing well? What is its reputation with the client? What is its cost performance? What skeletons are in its closets?

Audit agencies reveal dissatisfaction with the program, performance, counterparts, and contractor personnel. In fact, audit agencies are a highly skilled set of experts who measure poor performance when someone in Congress calls them in. When the whistle is blown, someone arrives from an authorized, investigatory body—somebody with the statutory authority to dig deeply into contractor records, subpoena witnesses, and obtain cooperation from the parties.

In every fiscal quarter, the audit agencies publish (through the Government Publishing Office, under a budget from Congress) a series of reports on all troublesome projects, and they are all troublesome to someone. The audit agencies determine whether the government got its money's worth from its contractors. These agencies start investigations after someone in the Department of Justice receives a worthwhile complaint. The trigger mechanism is usually the allegation of fraud, waste, or abuse.

The General Accounting Office

The General Accounting Office (GAO) works directly for Congress. Its job is to make scholarly and practical investigations of how government agencies spend their allocations. Reckless, irresponsible, and sloppy contractors are routinely eliminated by Congress because it is easier to wipe the slate clean than to straighten out gross mismanagement. Other contractors may endure inquisitions, hearings, testimony, and bad press, but survive the ordeal if they promise to reform. Still others are hassled and warned about infractions.

During a hearing, the firm's winning proposal is often read aloud and cited as an elaborate set of promises—promises that the contractor has not kept. The GAO thus shames the firm into doing a better job under threat of contract termination. In short, all these unflattering activities become readily accessible as GAO reports, to help us size up the competition.

The Defense Contract Audit Agency

The Defense Contract Audit Agency (DCAA) is empowered by statute to audit, critique, and withhold funds from contractors. The DCAA, however, is devoted only to the Defense Department—it covers no civilian agencies. The DCAA inspects, audits, and interprets contract performance in ways that stress a total, devoted work ethic. This strict, regulated view frowns on marketing, stresses flawless performance, and demands accountability for every Federal dollar spent. Monitoring time sheets, counting employees entering and leaving work at starting and quitting time, checking employee punch cards, making unannounced inspections, and checking desks are all part of its routine. It uncovers a contractor's avoidance of the government rules, and rectifies shortcomings. The DCAA is listed on all defense contracts; no wrongdoing has to trigger its normal scrutiny.

However, the Defense Department and defense-related agencies may request increased DCAA actions to hint to contractors that there is a performance problem. Like GAO reports, DCAA reports indicate that a contractor is performing poorly. Unfavorable DCAA and GAO reports affect a contractor's award fee where a CPAF *(cost-plus-award-fee)* contract was let, and such reports are also grounds for releasing a contract for rebidding at the first opportunity.

In very rare cases, GAO and DCAA reports lead to outright cancellation of work via a clause in the Federal Acquisition Regulations entitled "Termination for Convenience of the Government." Such terminations equate firing an insubordinate or unruly employee, and reflect a general lack of skill, tact, responsibility, or talent. The terminations are part of public record, and become a part of a firm's similar experience, albeit a page that most would sooner forget in their forthcoming proposals.

© 1996 by Holbrook & Kellogg, Inc.

The Inspector General

The Inspector General (IG) is the last and most serious inspector. Its visits are triggered by formal congressional, DCAA, or GAO inquiries, coupled with distressed pleas from the agencies themselves to send field investigators. The IG digs through records, seizes files, and conducts comprehensive interrogations of contractor personnel. Its interest in a firm or project is only one step away from the Justice Department, FBI, or state and local police investigations.

It is important to note that the GAO, DCAA, and IG operate under nonjudicial authority; that is, their advice to contractors and agencies is legally nonbinding. However, contracting officers recognize the audit agencies for the watchdogs they are. It is risky business for a government contract shop to award a new contract to a firm with bad marks at the IG, DCAA, or GAO. Contractors are expected to correct each perceived flaw or shortcoming.

Performance — The Key to Success

These audit agencies are the weather vanes of contract performance. Their reports, which criticize specific projects, are on file for public scrutiny. Complete, hard evidence is neither shown nor required; the mere inference of shoddy work will provoke negative reports. As the audit agencies go, so go the contractors, because the ability of the DCAA, GAO, and IG to influence funding is well known, and well documented. Therefore, people who wish to write stellar proposals should have a substantial number of flawless (or nearly flawless) projects upon which to base their similar experience. In solicitations, similar experience in the subsequent proposals *is always asked for,* usually in this format:

- Title of contract and contract number;

- Name and phone number of the contracting officer;

- Name and phone number of contracting officer's technical representative;

- Agency, branch, and address;

- Brief scope of work;

- Contract period of performance; and

- Original cost *versus* final cost.

A careful scan and reference check of such data by source selection officials and contracting officers soon reveals a pattern of good, average, or poor performance. The risks of choosing a poor performer outweigh any promises that the contractor's proposal may make. Again, contracting officers risk their own careers when they select firms with problematic track records and tainted performance. Thus, audit agencies play an exceptionally important role in the credibility of firms.

While Congress, or rather political patronage, can aid in the initial selection of contractors, it cannot overcome the hard evidence of audit agencies. Therefore, it is necessary for proposing companies to support their performance promises with solid, dependable information, much like a job applicant who supplies character references. The audit agencies thus greatly influence our clients' perception of us.

3.1.3 Public Sources of Data for Assessing the Opposition

The Federal system and most state governments publish budgetary data explaining where they spend their funds. Public data accessible at the Government Printing Office (GPO) and agency publication offices include:

- The budget of the United States;

- The Department of Defense budget for that fiscal year for the Army, Navy, and Air Force;

- Other federal agency budgets for that fiscal year, such as the plans for the procurements of the Departments of Energy and Education, NASA budget books, and Nuclear Regulatory Commission booklets; and

- The *Congressional Record*, *Congressional Digest*, and hearings records.

It is essential that company marketing personnel identify and acquire these budgets. Secondary budgets then describe which contractors are currently engaged in fulfilling current contract obligations by dollar value, period of performance, and other key data elements such as the work for which each firm is responsible.

Next, the firm should search for new information among business and technology journals where raw data, doctrine articles, and new projects are often announced. These include:

- *Business Week*;

- *Fortune*;

- *U.S. News and World Report*;

- *Aviation & Space Week*;

- *The Wall Street Journal*;

- *Harvard Business Review*;

- Papers and books published by the Brookings Institute;

- *The Washington Post*;

- *Army Times*;

- *Navy Times*;

- *Air Force Times*;

- Journals specific to the concerns of particular industries, such as *Signal, Electronic News,* and *Journal of American Defense Preparedness Association*; and

- *Commerce Business Daily* (The Awards Section shows you who is winning, what their price was, and what the period of performance will be.).

An employee should be assigned to receive, scan, and route publications through the firm's Technical Library and out to interested readers. A reading file from outside should complement the library with agency reports that have been obtained and are pertinent to forthcoming procurements. A clipping service can usually be justified based on a relatively large number of articles—clipping services usually do a satisfactory job for a fraction of the price a full-time employee would cost.

3.1.4 Private Sources of Data for Assessing the Competition

It is not sufficient for public information alone to fuel a firm's marketing activities. Private, special data must be brought out of agencies in order to provide substantive marketing intelligence. These data include:

- Allied parties who want to change jobs, be promoted, or effect some change within their area;

- Consultants within industry;

- Subcontractors who see key opportunities and pass along the news to a firm they feel is correctly positioned to use the information; and

Many firms now take extensive steps to preclude the loss of special private data by cutting off its flow. The steps usually are the following:

- Keeping plans and schedules out of the hands of people who are not trusted;

- Eliminating tours to visitors, vendors, and guests;

- Limiting personnel to a "need to know" posture;

- Selling technology carefully so that it will not be exploited by opponents;

- Staying aware of front companies who buy goods, then ship them to the tear-down shop of their competitors;

- Staying alert to pilferage, theft, and misuse of sensitive information; and

- Advising technical and business managers to scrutinize any item of information that knowingly leaves the firm.

Nevertheless, a wary practitioner of business intelligence has ways to get into the competitor's private data base. The most effective means to get private data are from interviews with past employees of other firms. The psychology of interviewing such people is best left to specialists in industrial relations.

3.2 CAPABILITY AND QUALIFICATION STATEMENTS

We learn little from capability and qualification statements due to their public relations nature. However, we should obtain the most current statements each year from our competitors, and perform these steps:

- List all projects and contracts shown by the opponent's firm;

- List all corporate strengths shown;

- Compare these strengths and credentials to the industry at large;

- Contrast the opponent's skills and weaknesses to our own; and

- Draw up a trade-off matrix for planning where and how to team with, or compete against *each major competitor*.

This trade-off matrix should identify (1) revenue (ours *versus* theirs); (2) the number of employees; (3) claims to fame; (4) skills specialties, such as ILS, oil and gas engineering, or space station design; and (5) items about which the firm brags.

We should pay attention to our own capability and qualification statements to limit valuable data. A wise practice is to publish such statements once a year, and omit new technologies and truly unique accomplishments. Press releases should be equally judicious.

3.3 CONFERENCES, ASSOCIATIONS, PROFESSIONAL SOCIETIES, AND TECHNICAL JOURNALS

Every year, a professional intermingling of information in each industry takes place, in which materials are discussed and disclosed in order to capture or hold new areas of the marketplace. The meeting places are most usually trade shows where clients come to see, touch, and buy new technology. These conferences, trade associations, and conventions are sponsored by professional associations like the Institute of Electrical and Electronics Engineers (IEEE), the American Defense Association, and others who have broad industry goals.

These shows are conducted and budgeted *individually* by firms who want *the credit* for technical accomplishments. Similarly, papers appearing in journals and magazines, under corporate sponsorship, are biased due to the vested interest of who is paying the bills. In fact, most reprinted journal articles are little more than glorified commercials.

Booths of persuasive literature, films, hospitality suites, and job interviews are a part of any well-run association meeting. Resumes are collected in the thousands. Firms line up their marketing plans for three to six months in the future. Reprinted articles glorifying various firms are handed out. In essence, the released data are controlled, sifted, and filtered, and yet are the rudimentary elements of a network, raw and ready to be developed for future marketing and proposal uses, and a means to size up the competition.

3.4 INDEPENDENT RESEARCH AND DEVELOPMENT (IR&D) PROJECTS

If we identify an ideal spot in the marketplace, and assess the competition as being either inept or handicapped in that area, then it is time to spend *independent research and development* (IR&D) money to exploit that gap. The IR&D money is a fund the Internal Revenue Service (IRS) returns to larger contractors, so that they can research specific areas in which the government believes there is a future. Just how specific and how noncommercial such areas must be is a matter of opinion.

Major corporations sometimes fund their proposal development groups out of IR&D funds as a means to substantiate the use of writing and editing personnel in the pursuit of technology. IR&D funds are also properly used to pay for the personnel, test facilities, materials, and consultants involved in the research.

Because these projects come from public funds (like contracts) they come under the scrutiny of Congress, audit agencies, and other data base generating organizations. In brief, *applied research* is often achieved with IR&D money. To know what IR&D

projects are underway at the opposition is to know what the competitors cherish, and what they plan to do next. Therefore, we should investigate the following areas:

- Where the opposition spends its IR&D money, and what results it is getting;

- How much is invested by the firm and in what ways—laboratories, test programs, *et cetera;*

- What our own firm is doing, if anything, to offset the opponent's thrust, to anticipate breakthroughs, and to get competitive results; and

- In brief, what the competition is doing with public funds, which will assist it in taking a dominant place in the marketplace.

3.5 FEASIBILITY STUDIES—PREPARATION OR INTERCEPTION?

IR&D projects are important because they will become feasibility studies, which in turn justify new projects. A *feasibility study is* a special kind of proposal that is unsolicited, independent, and unique. The feasibility study is the seed or nucleus of a fresh, new idea, an idea around which new business can be built. There are a number of reasons to prepare a steady flow of feasibility studies for clients:

- Unsolicited proposals and feasibility studies make bold statements about what industry feels is sorely needed.

- The studies are unencumbered by formal procurement restraints; that is, they do not have to meet RFP criteria or the criteria of the Federal Acquisition Regulation.

- These so-called "white papers" spur clients' interest in R&D, and may become contracts funded for their own intrinsic values.

- They become part of a *capture strategy* to establish the firm as *the technical innovator.*

- They influence the IEEE and the American National Standards Institute (ANSI) to start new committees to investigate the idea.

- They offset competitors' more conventional approaches with their special, unique appeal.

- They justify constant visits to the client's office, and get constant exposure.

- They influence the customer's judgment as to what ought to go into the scopes of work of future procurements.

- They provide all the key elements to link the RFP to our firm alone.

Let us consider a trial case in which IR&D money is spent to produce the applied research and feasibility study for a new computer chip that performs faster, costs less, and is produced and tested more easily than anything else in the marketplace. Sympathetic clients investigate the product and find that it yields 80 percent of the promised performance, according to independent testing authorities. Gradually, industry-wide acceptance will come, but in the meantime, who is perceived as the industry's front-runner? Whose name appears in journals and elsewhere in the press? And, most of all, who gets the R&D contracts to prove the complete concept?

The feasibility study is a breakthrough device, a secret weapon in the arsenal of the contractor that has the skill, talent, and energy privately to investigate and pay for unconventional, unorthodox ideas that just may work. Clients recognize these efforts as:

- Creative, inventive, ingenious, and exciting;

- A break from the monotony of traditionally acquired items; and

- A means to circumvent the narrow bureaucracy, and the prohibitions encountered when a procurement makes its path along various agency offices.

Lastly, and most importantly, feasibility studies become the cornerstones upon which requests for proposals are generated. Once procurement shops and contracting officers realize that old techniques and products have been made obsolete, they begin to shape new scopes of work to catch up. Competitors are then presumed to have inferior ideas.

The firm with the technologically advanced chip, therefore, has established itself in the marketplace with known, physical, performance characteristics. Procurement shops and contracting officers can, therefore, *specify* certain performance traits, such as certain speed performance, certain test performance, certain production capabilities for ensuring quantity, and cost performance parameters, seemingly without bias or forethought, and yet be assured of one certain product or service.

These characteristics become evaluation criteria for procurements, and serve to eliminate competing designs or ideas. Put simply, only one firm can satisfy the need. That firm wins according to the supposedly unbiased RFP criteria which, in fact, were at one time only an idea in the mind of an engineer or planner. Despite the semblance of competitiveness, the feasibility study had been successful in "wiring" the procurement with preselected information known to a specific few.

IR&D funds are quite frequently used to locate and fulfill such opportunities. This step *intercepts* the solicitation before it becomes the fully entrapped RFP that is advertised in the *Commerce Business Daily* or elsewhere, and negates any influence from those firms that wish to imitate the invention or breakthrough that originated in a feasibility study. To *intercept* the procurement is the goal of the feasibility study—to find, identify, and specify goods or services above and beyond the capabilities of the competition. These are the prudent steps of a firm that wishes to form an independent, self-sustaining base.

3.5.1 Executive Summaries and Feasibility Studies

The *executive summaries* of feasibility studies must be far more than those of proposals due to the argumentation needed to justify the feasibility study. Let us contrast the proposal to the feasibility study. The proposal:

- Is a known commodity based on a Federal RFP;

- Has contracting officers' support;

- Is funded;

- Is based on traditional approaches; and

- Has established cost parameters.

Whereas the feasibility study:

- Is an unknown;

- Has, as yet, no sponsorship beyond its own firm;

- Is not funded by any agency for development or production;

- Is experimental and may be unorthodox; and

- Does not have well-defined costs for numerous applications.

The executive summary of such a study has to give full justification of all factors involved in the concept. The summary must do more than just agree to an RFP. It must detail facts and opinions to move its argument flawlessly to a conclusion. Without an extensive executive statement, the feasibility study is just a collection of data bound together under one cover. Any omission can result in the negation of the argument, and every serious counterargument must be dealt with in enough detail to negate it.

3.5.2 Technical Appendices to Feasibility Studies

The study relies on its executive summary, text, and examples (appendices) to provide evidence or proof. The tests, procedures, results, logs, and other verifiable additions must be logical, clear, and concise in order to convince readers that something of value occurred. No raw data will suffice, nor will unsubstantiated opinions. No outspoken, belligerent rhetoric will satisfy the skeptical reader or the technical critic. Records shown as part of the study should therefore be direct, clear, straightforward, and accurate. They should support the executive summary, and can become the grounds for later standards and specifications related to the core technology.

The appendix plays a special role in technical persuasion because it is both raw and shaped data at the same time. The appendix is raw in that it captures the whole experience of research, errors, and accidents, yet is shaped by its framework, and its introductory and explanatory materials.

3.6 PRELIMINARY RFPs

It is a standard facet of business development to try to put as much of the firm's own ideas into RFPs as possible. By 1983 or so, Federal solicitations for very large procurements were released twice: once preliminarily, and once for actual competition. Thus, the front-runner has two opportunities to intercept the RFP, affect it, and slant its requirements in a winning way. The preliminary RFP is a version released by procurement shops in order to receive constructive feedback and to help eliminate the growing number of protests registered by losing contractors.

The preliminary RFP should be subjected to the same scrutiny that the actual RFP would receive. The front-runner should miss no opportunity to infuse the RFP with technology, selected kinds of required similar experience, and special personnel needs.

Opponents may be sized up by counting how many preliminary RFPs they pursue and attempt to influence. Also, they can measure *our progress* by spotting RFPs methodically aimed at our own industry. A key to participating successfully in the preliminary RFP process is to contribute winning ideas succinctly without their being readily identifiable as *our contributions*. In many cases, we can quote specialized professional standards to substantiate why recommendations were made, without compromising the true logic. The preliminary RFP has these contractual obligations:

- To aim the RFP at industry at large;

- To highlight key criteria that could be protested, and ask for recommendations;

- To avoid any semblance of a "wire"; and

- To include items inadvertently omitted in first drafts.

However, the wary contractor sees the preliminary RFP as:

- A source of special data, which must be evaluated;

- A means to identify a No-Bid early, or tell the agency about perceived flaws;

- A means to identify protestable RFPs, and obtain changes in potentially "wired" solicitations; and

- A way to improve the RFP equally for all bidders.

If an RFP is important enough for an agency to release early, and it fits your business plan, then that RFP deserves full and meaningful scrutiny.

3.6.1 Advance Work on the RFP

Whenever possible, it is advisable to study each RFP as early as possible. The old RFP (from FOIA) can then be used to plan the "win strategy" ahead of the other interested parties. At the heart of a procurement is its scope of work, instructions, and evaluation criteria. If these pieces can be anticipated in advance of the scheduled date, they will suffice for a simulated proposal. The date the RFP is actually released for mailing is almost too late to start.

3.6.2 Simulated RFPs

In assessing the competition, the *simulated RFP is* often left undone or incomplete. However, simulating RFPs has a number of advantageous aspects:

- We are ready to respond to most RFP questions;

- If the RFP is delayed for one to two months, we gain a significant edge on the competition and we have a strawman or stalking horse with which to war-game; and

- Our first draft will be much better than those of firms that did not simulate the RFP criteria (and began writing only upon RFP release).

Rating the Client's Priorities

The most practical way to begin simulation or straw man preparation is to rate the client's priorities, articulated in the clearest business and technical terms. For example, if the customer's first priority is a management structure to contain and administer various technical offices, the simulated RFP should focus on management, not on technology. If the client is embarking on an intense R&D effort, then the focus would be on hands-on technology, with the management structure in the background. Thus, the engineering effort goes ahead as planned, unrestrained by unnecessary constraints.

White Papers and Drafts

White papers (by us and others) will outline what the client has been shown. If we add these data *to the client's overall response,* we should obtain a draft outline pointing to three or four chief matters. Again as an example, let us say the client is an architect and engineer office looking for contractors to build a facility which uses new, but not totally satisfactory, technology. Various articles were published, white papers were given, and IR&D funds were spent on specific plans, specifications, and drawings. The client has faith in 75 percent of what has been shown. However, the client has some nagging reservations about the reliability of the new methods of construction, the maintenance of the facility, and the long-term costs.

Any of these issues could cause the client considerable discomfort from its own management, from Congress (if the client is Federal), and from private or public oversight agencies. To safeguard the client and assure him of little or no risk, we should make the simulated outline and draft materials coordinate major factors and subordinate minor factors—which is what the final proposal should do. The outline should be optimistic about the new areas without overemphasizing them. One such outline follows.

I. **Architect and Engineer Services for XYZ Thermal Plant**
   A. Accepted services, experience, expertise
   B. New techniques discussed, explained, ameliorated
   C. Accepted hybrid services, buffeting new and old ideas

II. **Construction Schedule and Milestones**
   A. Accepted progress measurement
   B. New approach, impact ameliorated
   C. Accepted hybrid approach with schedule

III. **Inspection and Quality Control Provisions**
   A. Accepted QA techniques—hardware
   B. Accepted QC techniques—hardware
   C. Accepted QE techniques—software

If the opportunity presents itself to show the draft or simulated RFP to friendly parties at the client's offices, then we should do so. If the concepts can be infused into the actual RFP, it can put competitors at a severe disadvantage. If the client has closed its doors to outside ideas, then we have prepared ourselves for tackling the real RFP. Employees should then do the following:

- Write drafts to each major heading and break their parts down into subheadings;

- Study ANSI, IEEE, and other code and regulatory data bases to get the probable requirements, specifications, and standards of the industry;

- Provide independent schedules to create a time line for each activity implicit to the design concept;

- Use (or at least find and interview) those consultants at the forefront of the new technology at issue; and

- Deal with and overcome surprises in the RFP.

Goals of the Simulated RFP

The writing should be a draft only; no full-scale effort should be exerted to flesh out the document as though it were the actual RFP. (In fact, some experts on proposal preparation frown on writing *at all* until the actual RFP is in hand.) The simulated RFP has important goals in mind:

1. What gaps do we still have in technology?

- No knowledge of a new ANSI specification.

- No experience in testing a new process.

- No backup papers on where a concept emerged.

- No appreciation of who is the industry's expert.

2. **What gaps do we still have in personnel?**

- Who is on staff *today* who is competent to be the project manager for this job if it were activated *today?*

- Who can support that manager with hard technology that the client respects? Who has worked with this client whom the client would respect?

- Who can we recruit to fill the gaps?

3. **What gaps do we have in experience?**

- What current or recent projects are most similar and will yield the most results?

- What IR&D efforts show *allied results* we can use to persuade the client of our proper alignment?

- Can we embark immediately on a task order, subcontract, or purchase order with another industry leader *to get indisputable experience* on the job? A study, consulting opinion, or break-even task on the new concept will suffice to gain insight into the innermost workings of a project.

- Can we fill our gaps in experience through *teaming arrangements* with other firms in a contractor-subcontractor role? Can we buy or trade expertise to improve our market position for the procurement?

4. **Whom do we have to beat to win?**

In short, the simulated RFP produces good questions to get us ready to write the winning proposal. When we can answer all the questions given above in the affirmative, then we are ready to compete.

3.7 GO-NO GO STRATEGIES

The ultimate goal of sizing up the competition is to be well enough versed to bid alone, to form a competitive team, or to walk away and no-bid a losing opportunity. It is a hallmark of successful firms that they no-bid frequently, reserving their energy

© 1996 by Holbrook & Kellogg, Inc.

and funds for genuine can-win situations. Below, we discuss how to utilize your well earned database of information, betting on some winning RFPs and eliminating the rest.

The Competitor's Viewpoint

Many organizations think of *go-no go* decisions as independent, separable considerations that reflect only a one-time value judgment to stop or start a proposal. That is a faulty view. The go-no go decision is just one more way of sizing up the competition. Basic techniques are involved in an assessment of whether to bid or propose. These fail-safe points are:

- The opponents' declared strengths in *our* public data base;

- The opponents' strengths in Congress;

- The opponents' reputation among the audit agencies (and whether they have skeletons in their contractual closets);

- The opponents' strengths *with the client;*

- Knowledge obtained by our own intelligence gathering network;

- Our synopses of the capabilities and qualifications of others;

- Data that we have gleaned from conferences, professional associations, professional societies, and technical journal articles;

- Our knowledge of the competition's IR&D projects, feasibility studies, and white papers given to the client;

- What we have seen in preliminary RFPs makes us believe that (a) we are ahead, (b) others are ahead, or (c) it is not "wired" for anyone; and

- Our simulated RFP shows considerable advantages for us already (or liabilities, if it is slanted toward others).

In brief, go-no go decisions are made on the logical basis of how similar the work is, what we know about it, and whom we have who is competent to manage and design the effort. All the while, we *know* and *acknowledge* that we are either the front-runner, or else a lesser competitor with less expectation of success. The bottom line is not to bid on suspicious RFPs. The following methodology is aimed at an engineering approach to go-no go.

The Ten-Step Approach

For many American firms, the selection of which RFPs to bid on and propose is a vital part of business life. Therefore, we must be able to rapidly and accurately identify those RFPs that are most likely to win contracts, while eliminating those RFPs that will be troublesome, time-consuming, and likely to be lost in spite of the best efforts of the best team. To propose, or not to propose; that is the question.

Few companies have gone to the trouble of clarifying how they pick RFPs, and it is not unusual for managers to take an "I'll fly by the seat of my pants" attitude, trusting intuition and hearsay instead of factual, objective inputs. When a person considers how much of a company's time and money is tied up in an average technical proposal, it is not hard to make the judgment that all RFP activities should be methodically approached. While the following evaluation factors are thorough, they are not intended to be totally inclusive, but rather are given as a tool that can be finely tuned to the needs of a particular company.

The approach used in the ten steps is mathematical, in order to yield a straightforward evaluation to which all managerial, financial, and engineering personnel could contribute on a mutually understandable basis. Once such a common denominator is fixed upon, there can be no special or idiosyncratic interpretations of data that could confuse or blur an overall, broad view of whether or not to propose on a certain RFP package.

The participants of the team making such decisions should remember that they are contributing to a company-wide effort in which no single professional discipline predominates, which is of itself a means to omit esoteric criteria from the evaluation process. By using readily accessible mathematical steps for arriving at an evaluation, engineers and accountants do not have to develop lengthy and complex justifications for their calculations. Again, managerial time is saved by this process.

Thus, we can generally say that RFP evaluations can and should be objective, mathematically simple, and judged on one set of criteria.

All the following factors affect the project's success in roughly equivalent proportions; therefore, we rate each factor from 1 to 10. We do not use fractions other than one-half. Because the factors have been grouped under ten headings, the maximum score for a project's success would be an unlikely 100 percent. Several persons should evaluate the factors, and then compare their results in a meeting. Where wide differences of opinion are present, compromises could be made by averaging the diverging scores until one single score emerges.

Usually, projects that score 80 to 90 percent are good risks. Projects from 66 to 80 percent are fairly safe risks, but projects from 50 to 65 percent may merit deeper investigation prior to committing funds or people. Projects in the 30 to 50 percent range are probably long shots or dark horses, but may be worth the time and effort of a bid or proposal in order to "sound out" the market, ascertain the competition, or get part of an award. (See Table 3.1, an excellent checklist to use to thoroughly evaluate an RFP.)

Table 3.1
Evaluation Factors for RFPs

Factor	Issues	Score*
1. Experience	How much experience does this agency or company have? Has our company had more or less satisfactory dealings with this agency or company before? Did it meet our expectations in liaison? Were billings paid on a timely basis? Were there differences in philosophies or personalities on the job?	
2. Reputation	What is the reputation of this agency or company? If this agency or company has not done business with our company before, what can we ascertain about it from the business community at large? Does it have a good reputation among its stockholders? What is its Dun and Bradstreet rating? Has this firm been in the RFP area long? Is it competent to be proposing such a project? Does it have the managerial and financial strength to see it through completion?	
3. Personnel	How many technical people do we need to handle the project? Does this agency or company want a large crew, and will that handicap us elsewhere? Is the buyer willing to bear the cost of salaries of the specialists who are being requested? Is the proposed manager available for this job, and can the manager be depended on to see it to completion? Are other key personnel available? If not, can other projects be adjusted satisfactorily to staff this job?	
4. Context	What is the situation at the job site? Is this company or agency under pressure to perform? Is its proposed schedule unrealistic? Are there regional factors involved? If so, are we at a disadvantage?	
5. Competition	Who is the competition? Do we know from RFP lists, meetings, or from other sources, who will compete for the contract? What is our frank estimation of its managerial and technical ability? Is its staff capable of manning the project? Are we significantly ahead of or behind this company? (If it is very likely to win, we are entitled to only one or two points here.)	
6. Technical	What skills are required to complete the job successfully? Is the competitor more or less qualified than our firm? Are we more specialized or too general for the job? Is the competitor more specialized or general for the job? Is the competition stretching credibility here? Are we? Are there any new personnel at either company who could make a difference? Are any of our recent employees with a competitor now?	

Table 3.1
Evaluation Factors for RFPs

Factor	Issues	Score*
7. RFP Criteria	Has the competition ever won this contract before, or even won other contracts from this agency or company? Does the competition have an inside track with the buyer, or do we? Are the criteria "loaded" in somebody's favor? Are there odd types of software required? Are there odd manufacturing and test criteria? Are there strict, unusual personnel requirements? (These are key clues to a "loaded" RFP.)	
8. Contract Provisions	What type of contract is being offered by this agency or company? If the contract documents have any unusual provisions or new clauses that are beyond the scope of usual contracts, there may be additional risk involved which is not evident on the surface. Are there any dangerous or peculiar circumstances that may hinder our performance as the job is stated?	
9. Profit	What margin of profit is involved? Does the contract offer enough profit and incentive to commit a lot of personnel to this job for the years involved? Can we make additions to the contract at a later date which would add more profitability? What is the likelihood that the project will not be funded, or may be canceled?	
10. Impact	Will this contract add to our overall reputation? Are there environmental, political, or other connotations to this project that could give us some adverse publicity? If so, can it be offset, mitigated, or counterargued satisfactorily? If not, is the profit worth the risk?	
	Total	

*Score 0 - 1 for no data; 5 - 6 for average; 9 - 10 for excellence.

These, then, are the decisions that a company must weigh prior to committing itself to an extensive proposal effort. However, an intangible factor often enters: lobbyists' representation for the company. How extensive are the lobbyists' efforts in meeting with and convincing elected and appointed officials of the wisdom of a selection? This type of technical persuasion must also be managed to ensure better success.

3.8 ADJUSTMENTS TO STAY COMPETITIVE

A log should be maintained by each firm that lists each RFP ordered, each RFP bid proposed upon, and the win and loss tally. We should also note the number of proposals that we had to write to win one contract. Domination of the marketplace by one company usually indicates technical superiority or better cost performance. (See also Section 10.6, how to recover from losses positively or enjoy the euphoria of a victory constructively.)

In order to capitalize on victories and recover from losses, we must adjust the sources of errors to eliminate major, repetitive flaws such as poor approaches, high costs, and lack of quality management. Error-free corrective actions yield performance. The questions that the firms should continuously ask themselves are the following:

- Why did we win?

- Why did we lose?

- Why did we make the short list, and then lose in the final evaluation?

- Why did we fail to make even the short list?

- Did we make cost and technical mistakes?

A few important factors go into losing proposals, and of course, we have to *adjust* for each factor. Here are some of those flaws and their cures (See Table 3.2, a checklist for faulty proposals):

Table 3.2 Adjusting for Errors	
Errors	**Corrections**
Proposing to client who rejects our firm—that is, constant losses.	(1) Change the client's mind. (2) Do not bid on any of its RFPs. (3) Seek subs from victors.
Selected for short list but few or no awards.	(1) Team up with winners. (2) Bid on smaller contracts.
Spotty victories, some selected wins.	(1) Improve and intensify marketing. (2) Bid more selectively.

3.9 SUMMARY

The most frequent reason for losing is bad marketing intelligence; that is, a chronic inability to size up the competition. The most frequent reason for a victory is the exploitation of accurate marketing intelligence. Proposal teams and the firm's proposal development organization should make prompt, positive adjustments to trends, remembering all along when to team up with others, when to go it alone, and when, most importantly, not to bid. A no-bid is an affirmative, positive indication that a firm knows its marketplace, and better yet, those areas that are *not* its marketplace.

CHAPTER 4

Proposal Development Organizations

The proposal team is most effectively created from an in-house *proposal development organization* (PDO) from which the firm can choose many highly skilled professionals for each proposal effort. The proposal development organization should be comprised of a team of proposal experts who can routinely plan, schedule, outline, write, edit, and illustrate the document. PDO groups should be sufficiently staffed so they can provide the maximum benefits to the company without straining themselves to the breaking point.

The proposal team should be supplemented by subject matter experts, engineers, and consultants to write and manage selected parts of the proposals. A PDO should organize and support each proposal manager at an optimal level established by the corporation's top management.

Selection of the proposal manager is crucial to both proposal development organizations and the company at large. No individual is more important to a PDO and proposal team than the proposal manager. He or she is in charge of writing a successful proposal, and because the proposal is considered the product of a single, gifted mind, the proposal manager thus should be able to write and edit well.

Proposal specialists, and established and *ad hoc* PDOs, are in an excellent position to identify and recommend individuals for the job of the proposal manager. These are the characteristics of a strong proposal manager:

- Knows the RFP proposal process intimately;

- Knows the company's methods intimately;

- Understands the client's goals and aspirations;

- Is conscientious, exceptionally diligent, and thorough; and

- Knows how to use a proposal development organization optimally, and is especially adept at delegating nontechnical and proposal administrative matters to the deputy proposal manager (who is usually a PDO person).

The proposal manager and deputy proposal manager oversee, manage, plan, and schedule each aspect of the proposal's preparation. Figure 4.1 illustrates a typical, recommended proposal organization. It shows how the PDO group supports the proposal writing team and the associated Red, Blue, and Gray Teams (Section 11.1).

Because the proposal is a paper product, the proposal development organization's role is to assist the proposal manager in releasing quality written materials on schedule. A PDO is the right arm of the proposal manager for these prewriting activities in Stage 1:

- Helping with the proposal strategy;

- Assembling similar experience, resumes, and past, similar proposals;

- Cultivating the client further; and

- Evaluating the competition.

The actual writing begins at Stage 2, and the PDO is responsible for these activities:

- Managing plans and scheduling all efforts;

- Assigning the most talented employees to the most challenging areas;

- Analyzing the RFP, cover to cover;

- Giving clear, complete instructions to volume managers, section leaders, and individual contributors;

Proposal Manager

- **Deputy Proposal Manager**
  - PDO Proposal Management
    ▸ Proposal Plan
    ▸ Proposal Specialists
  - Editors & Illustrators, Production

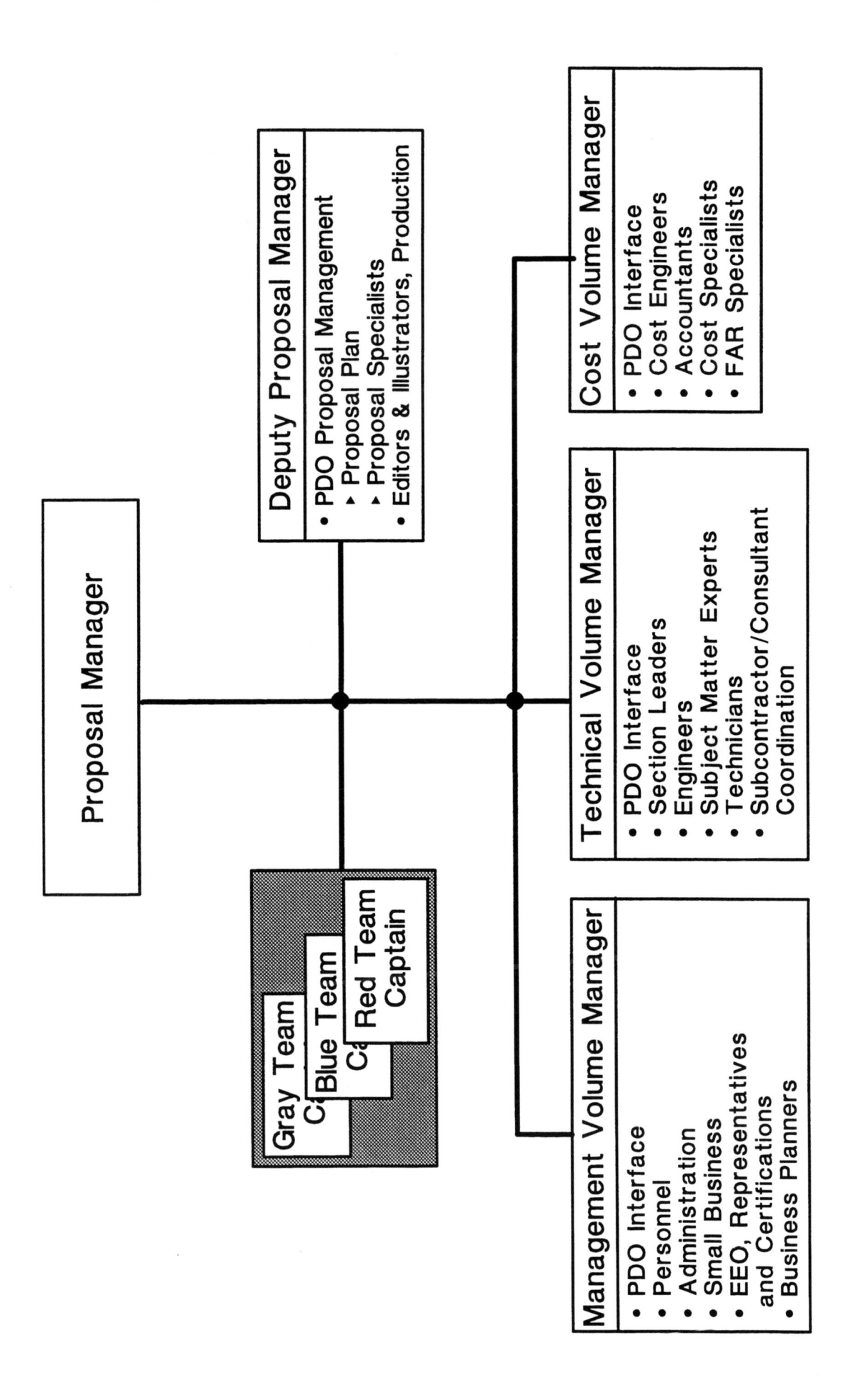

- Gray Team Captain
- Blue Team Captain
- Red Team Captain

- **Cost Volume Manager**
  - PDO Interface
  - Cost Engineers
  - Accountants
  - Cost Specialists
  - FAR Specialists

- **Technical Volume Manager**
  - PDO Interface
  - Section Leaders
  - Engineers
  - Subject Matter Experts
  - Technicians
  - Subcontractor/Consultant Coordination

- **Management Volume Manager**
  - PDO Interface
  - Personnel
  - Administration
  - Small Business
  - EEO, Representatives and Certifications
  - Business Planners

Figure 4.1
Proposal Development Organization Interfaces

- Reviewing progress in text and art, evaluating the quality and persuasiveness of the materials;

- Resolving disagreements between contributors;

- Creating, reviewing, and revising the proposal win strategy and its themes, sub-themes, and selling points;

- Coordinating the storyboard materials, drafts, issues releases, word processing, graphics, and reproduction; and

- Editing text and producing the art.

In short, the proposal development organization and the proposal manager actually manage by *editing*. The advantage of having an established PDO to rely on is that its staff will be acutely aware of the significance of imprecise responses, data unrelated to the RFP, and awkward engineering prose that needs revision. The firm's technical writers and editors can work on the proposal only if proposal writing is a routine, expected part of their jobs. Otherwise, they may work on the proposal leisurely as though it were a slowly paced document like a technical manual or deliverable report. The completed proposal should read as though it were written by a single author. This clarity and coherence can best be achieved by a technically expert program manager whose staff is guided by proposal development specialists.

4.1 PROPOSAL MANAGERS AND PROPOSAL DIRECTIVES

When the proposal manager and proposal development organization take charge of the proposal, the most straightforward means to notify the team of its collective and individual responsibilities is via the *proposal directive,* a typical example of which is shown in Figure 4.2 in memo format.

The proposal manager must initiate precise directives for the following:

- Notification to PDO of a major effort;

- RFP receipt and area assignment;

PROPOSAL DIRECTIVE NO. 001
(NAME) PROPOSAL

TO: (All contributors listed by name)

FROM: Proposal Manager (name)
Deputy Proposal Manager (name)

SUBJECT: Plan, Schedule, and Milestones for a Winning Proposal for (name) Proposal

Attached is an assignment of personnel to tasks. Your assigned portions are indicated per our discussions.

For technical matters refer to (name), Technical Volume; for costs, refer to (name), Manager, Cost Volume; for Management/Business, refer to (name), Management Volume. At our daily staff meetings, feel free to discuss any matter affecting us all.

Attachments: schedule, plan, milestones

Figure 4.2
Proposal Directive Memo and Schedule

© 1996 by Holbrook & Kellogg, Inc.

Schedule and Milestones				
Event	Done By	Due	Finished
Receive RFP			
Analyze RFP and Develop Checklists			
Submit Management Sections			
Adjust Outline and Strategy			
Finalize Offering			
Write Overall Theme Statements			
Prepare and Review Assignments			
Revise Management Sections per RFP			
Draft Cost Sections			
Hold Kick-off Meeting			
Write Technical Section Themes			
Review and Approve Technical Themes			
Draft Technical Sections			
Draft Technical Visuals			
Submit Boilerplate to Technical Publications			
Review and Approve Management Sections			
Input Management Sections			
Edit Management Sections			
Submit Cost Sections			
Check Certifications			
Correct Management Sections			
Review and Approve Cost Sections			
Input Cost Sections			

© 1996 by Holbrook & Kellogg, Inc.

Schedule and Milestones			
Event	Done By	Due	Finished
Review and Approve Technical Visuals			
Management Review of Management Proposals			
Receive/Review Sub's Proposal			
Submit Technical Sections			
Edit Cost Sections			
Review and Approve Technical Sections			
Return Sub's Proposal for Revision			
Correct Cost Sections			
Management Review of Cost Sections			
Input Technical Sections			
Revise Management Sections per Review			
Edit Technical Sections			
Correct Management Sections			
Revise Cost Sections per Review			
Draft Executive Summary			
Correct Cost Sections			
Correct Technical Sections			
Receive Revised Sub's Proposal			
Submit Proposal to Red Team			
Hold Red Team Review			
Produce Final Draft			
Revise Technical Sections			
Identify Proprietary Material			

© 1996 by Holbrook & Kellogg, Inc.

Schedule and Milestones			
Event	Done By	Due	Finished
Produce Table of Contents			
Final Edit Technical Sections			
Produce Compliance Matrix			
Input Corrections and Proofread			
Write Cover Letter			
Print and Bind Proposal			
Submit Proposal			

Figure 4.2
Proposal Directive Memo and Schedule (continued)

- Kick-off meeting;

- Section and subsection assignment;

- Subsection assignment of subcontractors or consultants;

- Blue Team action;

- Red Team action; and

- Final copy of the proposal, and publishing and shipping arrangements.

Sometimes, a few contributors fail to meet their obligations on schedule with the quality that we desire. Tardy contributors often blame incoherent directives and memos to avoid being held responsible for slipping the delicate proposal schedule. Therefore, it is better to direct a strict, tight schedule than a loose one. The tight schedule is easier to monitor but harder to manage because it demands that so much periodic text be produced on time, in quantity, and with quality. When the proposal manager and deputy proposal manager agree that pitfalls are occurring, it is essential to issue an intermediate proposal directive, which alerts people to sore spots, assigns new people, and relieves unsatisfactory contributors of their assignments.

© 1996 by Holbrook & Kellogg, Inc.

Upper management will oversee a proposal team's progress by the flow of proposal directives, so it is important to make these special memos succinct, and to demand the resources needed to succeed. In addition to the "official" directive process, senior managers informally walk in and see if everybody is gainfully dedicated to real work, and ask how the PDO is achieving its goals. The proposal manager and deputy proposal manager need to be one or two steps ahead of these checks by making sure that people are working as directed. It is prudent to replace bad workers immediately. A few unproductive or uncooperative people can wreak havoc on a proposal in a few weeks if such behavior is tolerated.

Again, directives must be written, assigned individually by name, and scheduled to succeed on a master timetable, like a PERT chart used by a project manager and deputy to control each element of work. The proposal directives create a management audit trail by which upper management, PDOs, the Red Team, and others can see real progress taking shape.

4.2 A DEDICATED, PERMANENT PROPOSAL DEVELOPMENT ORGANIZATION

The proposal development organization should be a dedicated, permanent part of the company's marketing, bid, and proposal team. A PDO provides a central focus to values such as writing skills, editing skills, archives of similar projects and proposals, and accelerated RFP responses. Here are some advantages of a permanent PDO:

- The PDO is there when we need it;

- The PDO reminds the company of wins and losses;

- It specializes in written products; and

- The PDO team routinely faces and solves proposal crises.

However, a permanent PDO is rather expensive to staff and run, and its costs are overhead, and therefore not billable to the client. Nevertheless, a permanent PDO is a good investment, for the following reasons:

© 1996 by Holbrook & Kellogg, Inc.

- Proposal development organizations are geared for a fast pace, and are ready to respond to the proposal manager's accelerated schedule;

- PDOs can rapidly familiarize engineers and technicians with proposal techniques;

- PDO people know the ropes—they lose no time and waste no energy; and

- The alternative to a PDO is to staff each new proposal with new people of mixed quality—a risky venture indeed.

Thus, it is evident that a dedicated, consistently funded PDO is a major enhancement to bidding, proposing, and marketing success.

4.2.1 PDO Manpower

To provide enough manpower for annual proposal efforts, we should compare last year's workload to this year's workload. We tally how many of the following employees were used, and when: PDO managers, deputy proposal managers, proposal administrators, proposal specialists, editors, and illustrators.

To illustrate the use of manpower for developing proposals, let us consider the hypothetical Company A. Every year, Company A writes 36 full-blown proposals, 24 of which take 60 days to write, and 12 of which take 45 days to write. Company A has a dedicated PDO team, which is supplemented by consultants and part-time, hourly help. It has six qualified PDO managers. Each manager undertakes four major proposals a year, one a quarter, in addition to other duties on smaller proposals. Therefore, these six employees can complete the annual workload and render some aid to the smaller, less significant proposals.

Company A has 12 supplementary PDO persons, technical writers, and graphics illustrators paired to each proposal manager. Thus, every team is comprised of three people, who are physically located together as a concentrated work force. The PDO groups' combined win ratio is 25 to 30 percent for the past two years.

The management of Company A wants a 40 to 50 percent win ratio for the expended budget, but is dissatisfied with the hired consultants. Most proposals of Company A reach the best and final evaluation (they come in first or second out of several proposals), but some proposals are flatly rejected, which embarrasses Company A very much. Of the proposals reaching the best and final evaluation, 66 percent of these were rejected because of their cost.

These factors seem to indicate that the PDO staff is at or exceeding full work capacity needed routinely to write successful proposals. Lack of cost trimming on the technical side also is an apparent source of losses. It may be that PDO persons, the technical team, and cost people are concentrating too much of their efforts on preparing the original proposal with nothing left to expend on subsequent steps. In other words, while the PDO and others are gainfully used, the emphasis is falling too heavily on getting the proposal out, and then too lightly on correcting and clarifying the overall project for the best and finals. Teams that disperse after the original proposal is delivered are hard to reassemble.

Therefore, Company A must concentrate more on the postsubmittal phases. It should add a seventh PDO team of three persons to take over selected efforts, from start to finish. The members of each proposal team should start working together as soon as possible, because the sooner they feel cohesive and motivated, the better.

To pare down the budget, Company A can hire new cost people, and point out to all the teams that many of its proposals were defeated by lower priced offers. The new, enlarged staff is tested for six to nine months to evaluate its win ratio, ease of management, and to analyze any Red Team criticism.

Smaller companies that cannot afford a large PDO should try to staff each proposal thoroughly enough to ensure a win. However, an understaffed company should know enough to cast a "no bid."

4.3 PDO ALTERNATIVES FOR SMALL AND MEDIUM-SIZED FIRMS (*AD HOC* ORGANIZATIONS)

Small firms need to cultivate their PDO functions even more carefully than the larger firms do, because their resources, time, and people are more limited. It is imperative that a small firm select a proposal manager who is technically competent and understands the proposal process, because a weak manager kills not only everyone else's motivation, but may kill the proposal as well.

An *ad hoc* PDO person, such as the deputy proposal manager, should be available all the time to assist the manager whenever necessary. Consultants can be very useful and cost-effective for small firms, but they have to be competent, loyal, familiar with Red Team tactics, and not too expensive.

Management must allow a stable team to produce the proposal in a businesslike manner. Therefore, management must refrain from interrupting the project, making double or triple assignments, and criticizing the product prematurely.

By otherwise copying a PDO group, the small firm can improve its success ratio. Most small firms know their clients exceptionally well, usually personally. Therefore, the lack of unity between marketing and proposal groups, which afflicts larger firms, does not usually occur in small businesses. Another advantage of the small firm is that its PDO group can easily cover the firm in its entirety.

4.4 LEAST FAVORABLE CHOICE

Many small and medium-sized firms do not have a PDO or suitable substitute for their proposal activities. They rely on putting teams of strangers together on the spot, counting on a fusion of good will, loyalty, and technical skill to effect a polished end product. Sometimes this procedure works, when all the personalities mesh and management directives go as planned.

Most of the time, however, the people assembled in this way work independently, disjointedly, and without a central purpose. Winning and losing is relatively random, with no real effort to make corrections. Firms can survive this way, but they have no guarantee of success or of future contracts.

A manager who wants his or her company to adopt a PDO can try the following tactics. He or she can produce a white paper that justifies the costs of a PDO group by showing an increased win-loss ratio. The manager can constantly talk about those business colleagues who use a PDO arrangement successfully.

If the company still refuses to employ a PDO group, then at least it must do the following:

- Have a single vice president or senior manager supervising the proposals;

- Find the best technical consultants, and stay informed about what they are doing;

- Sponsor an annual proposal training seminar;

- Assemble Red Teams to carve up bad material; and

- Be prepared to manage a lot of crises.

However, there is no substitute for a well-staffed, well-organized proposal development group.

4.5 PROPOSAL COSTS: MEETING THE MILESTONES ON TIME AND ON BUDGET

Cost scheduling considerations are at the root of a successful long-range proposal effort. Budgets for proposal preparation deserve as much attention as the proposal's technical content, because financial considerations can and do constrain proposals. This section provides an approach to proposal cost and scheduling planning for effective management of the proposal process.

The preparation of proposals and feasibility studies cannot be accomplished without cost and schedule planning. In the worst cases, only the request for proposal deadline dictates time and money constraints. A more prudent, less haphazard approach is strict cost and schedule planning, like that used to protect management, so that each segment and the overall proposal gets appropriate, cost-conscious management attention.

4.5.1 Annual Plan

Proposal milestones must be planned at the highest level within each firm. An annual proposal activity plan, derived from the latest business development and marketing information, must outline expected levels of effort for the four quarters of the year. This budget is based on the salaries, benefits, travel, and support of all the employees working on proposals. The proposal activity plan must allocate enough money to cover these expenses.

A quarterly breakdown will reveal where work emphasis will fall throughout the year. Attractive, small contracts should get appropriate consideration, and risky larger ones should be studied carefully to determine when they will occur in a 12-month period. Activities from past years can be used to extrapolate and forecast resource requirements.

Once this estimate has been reviewed and determined to be an accurate summation of the situation, we must set aside funds for the proposal development organization and for individual efforts of responsible managers. The individual proposal is the level which disrupts the cost and schedule constraints most often. Pressing tasks are often construed as a rationalization to override the budget; the proposal team simply pulls out all the stops on expenditures.

Sometimes such actions are valid; sometimes they are not. However, throwing money at problems is ineffective project management, and it should be avoided in proposal preparation as well. Performance in proposal preparation is best measured by cost and schedule performance. In other words, it is imperative to run proposals and projects in the same manner.

Like a project, a proposal without a manager is in grave trouble; therefore, a proposal manager should be named as early as possible—perhaps as early as the annual proposal plan. The manager can then focus on the following considerations:

- What technical approach is likely to succeed;

- How much a particular proposal is likely to cost;

- How much a particular proposal will add in revenue;

- What a particular proposal will add in prestige; and

- What the risks and problems will be.

Budgets can then be matched to benefits, difficulties, and targets. For each major marketing area, we may write several key programs—but how many and by whom are important cost and schedule questions.

The high cost of proposals can then be managed for time constraints and projected budgets. The important rein on expenditures keeps the annual budget intact and preserves precious resources for contests yet to be encountered.

4.5.2 Cost Logic for Solicited Proposals

Four P's apply to cost and schedule management of proposal preparation: *Preplanning Prevents Poor Performance.* Long before an RFP appears, marketing and proposal managers should have scoped out the known and expected parts of a specific proposal; they do similar preliminary research on projects. One recommended sample proposal flow chart is shown in Figure 4.3, which attempts to display the time-budget relationship of the proposal.

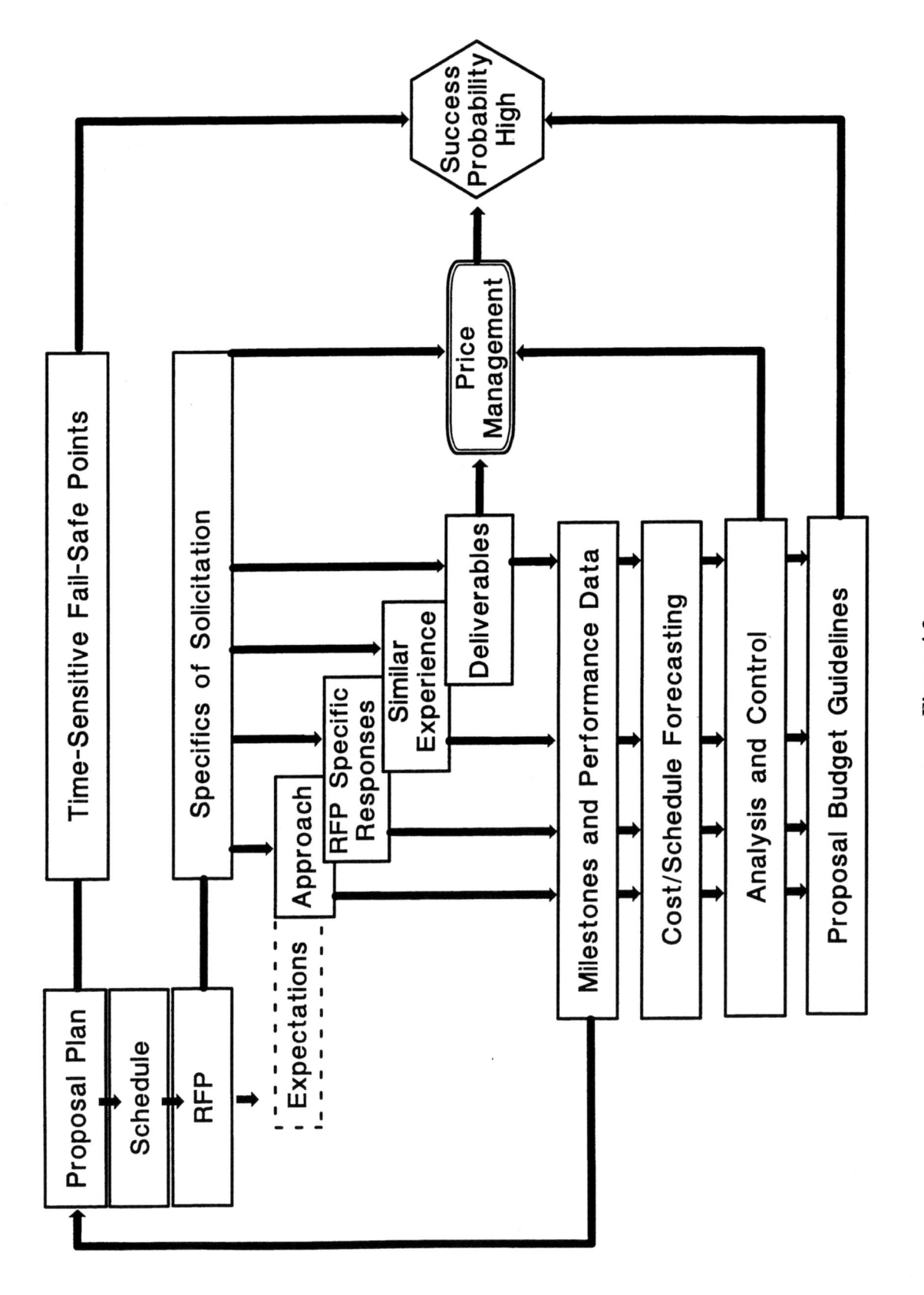

Figure 4.3
Proposal Flow Chart

The schedule, clients' known concerns, and time frame for response are then entered into this framework. Assignments of responsibilities should be made at this point (if they have not already been made as part of the annual plan) so that, when the actual RFP arrives, we are ready to insert real criteria into the categories of the flow chart. "Stem" supervisors should be named to head major divisions of responsibilities (like similar experience, resumes, technical approach, and price). Each stem supervisor further defines requirements for assigned areas, keeping costs as a prime consideration. As secondary overseers, the stem supervisor monitors cost performance as well as schedule and quality performance. This close management prevents crises and cost overruns.

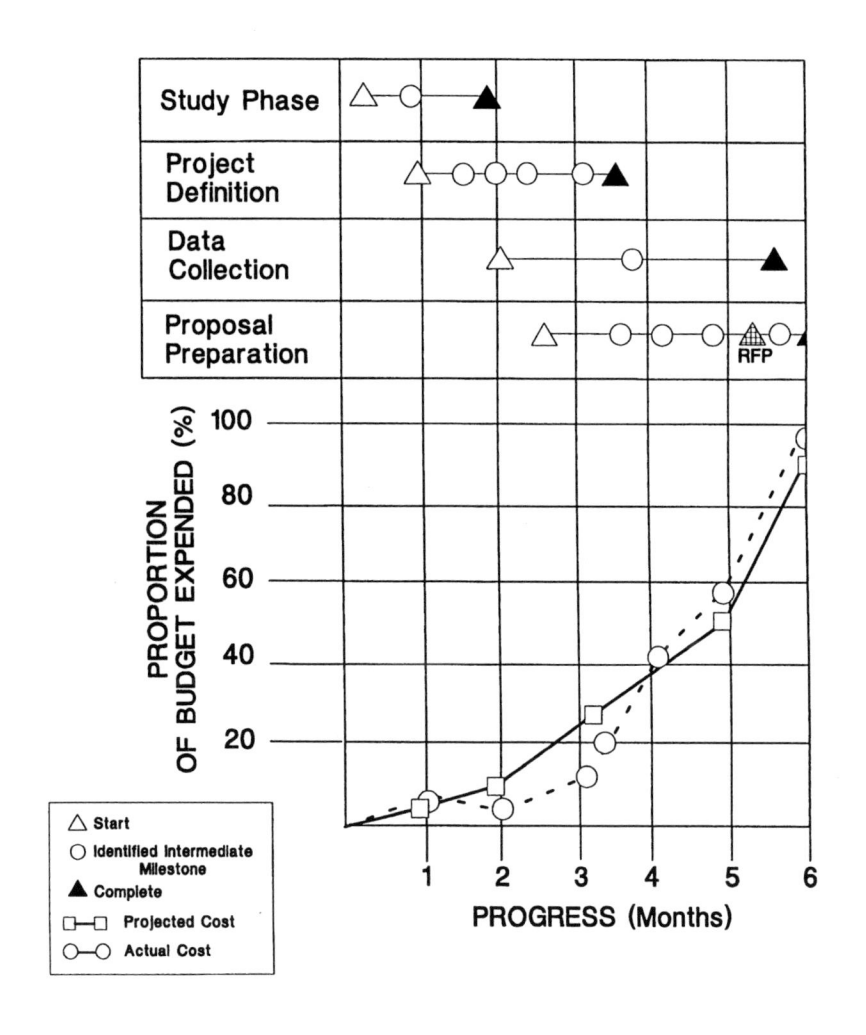

Figure 4.4
Proposal Cost and Schedule Plan

© 1996 by Holbrook & Kellogg, Inc.

Preplanning allows us to meet the milestones on time, with quality and on budget. Figure 4.4 shows a typical plan for achieving the desired schedule and budget goals. The timeline demonstrates that labor, time, and money are being tracked to assure the burn-rate of resources does not exceed the deadline. Of course, each company should tailor its own proposal flow charts and cost and schedule plans. Organizations cannot lavish efforts on proposals that have no potential for commensurate payback. We must make compromises in budget for any proposal not expected to result in a substantive contract—if we indeed write the proposal at all.

One industry standard states that proposal preparation may deserve a five percent expenditure of the proposed contract. Thus, a $100,000 contract may deserve a $5000 or $7000 proposal. Although this percent grows as proposal values increase, it must be considered a meaningful standard. The result of ignoring it is the so-called gold-plated proposal that burns up overhead budgets for proposal preparation.

The proposal preparation schedule should contain weekly, even daily, percentile goals for performance and cost. Certainly, a 30-day deadline means that at least 33 percent of the work must be done in each of the first three weeks, to allow for a week or so for management reviews. Proposal preparation is not a linear process, of course, and the schedule expands and compresses as the proposal management directs.

Supervisory reviews should occur daily, if not more often. This frequency of review prevents the last-minute chaos typical of proposals in which the costs and schedules are not managed carefully. As reports come in to managers and task supervisors, analyses and controls can be adjusted to address weaknesses, and resources diverted to areas where emphasis is needed. Cost is thus regulated.

As noted above, proposal and business development costs increase in proportion to the dollar value of a forthcoming contract. Time also increases because a year or more may be spent in anticipation of a single RFP or a family of related RFPs. The key to planning big, important procurements is to control costs by using only a few persons in the monitor role. Not only does this cut the cost of proposal preparation, but it also focuses the attention of these monitor people on long-range targets. Cost strategies can almost always be productive if we follow these guidelines:

- Segregate bid and proposal hours strictly from any clients' billable project charges (DCAA auditors look for abuses here);

- Whenever possible, ask only "overhead" professional or technical employees assigned to each proposal to carry out a task in order to avoid overtime charges;

- Minimize hourly employees' overtime;

- Limit the use of consultants to areas where their skills are genuinely unique;

- Get consultants to accept ceiling prices or firm, fixed price contracts; and

- Do not pad the schedule.

The basic rule for cost and schedule management is, "Do not solve problems by throwing money at them." Cost and schedule management will prevent the dire pitfalls that result from throwing money.

4.5.3 Cost Logic for Feasibility Studies

Proposals arising in response to an RFP are usually straightforward and manageable. Their relatives, feasibility studies, often branch out to the extent that they are sometimes imprecise, hard to manage, and costly.

In a feasibility study, the central issues are condensed into a single hypothesis which is then tried against all the competing and intellectually equal and hypothetically possible solutions. Various solutions eliminate themselves because of extensive time factors, extensive cost factors, or environmental, safety and health considerations. The feasibility study is written from a broad, expansive, comprehensive, nearly all-inclusive perspective so that the client has to appreciate the thoroughness and completeness of the examination. Equally comprehensive are the trade-offs to eliminate the options to a single, final solution to the singly articulated problem. Reducing the data to conclusions can be difficult, with a lot of trade-offs, and intellectual sword-play. Typically, some solutions prove themselves infeasible due to their use of outmoded technology, inferior equipment performance, too sophisticated a design or component, high costs, specification of materials that do not exist to date or have not been tested, or simply lack of an advocate or sponsor to fund that solution.

As infeasible avenues are eliminated, resources can be moved into productive areas, until the single most effective route is discovered. This route may be a synthesis of the two most promising routes. The in-house feasibility study can be expensive and time consuming, and must be strictly managed.

4.6 FINAL CONSIDERATIONS

At the end of each proposal or feasibility study effort, we should devise an evaluative, feedback loop to help prepare the following year's annual plan. Thus, cost and schedule planning comes full circle, with accepted project management techniques applied to the proposal and feasibility studies. The result is a proposed effort that will be successful over the long range. As we have said throughout Chapter 4, the proposal development organization is best suited to perform each of these cost-conscious actions.

As we have said elsewhere in the book, the goal of the company's proposal writing shop is to improve its operations year-round, focusing on the means to solve crises, resolve bottlenecks, and arrive at a quality product. In three decades of institutionalized, industrial proposal writing in the United States, the overwhelmingly most successful results have been yielded from permanent PDOs such as we discussed above, using many of the same techniques we advocate.

CHAPTER 5

Proposal Prewriting Activities

Writing winning proposals stems from being continuously prepared for most requests for proposals. This state of preparedness comes from continually monitoring the release of RFPs from clients and potential clients. Each decision to bid or not to bid aids the proposal development organization in assessing its performance ability by checking the following:

- The client's attitude of whether our company is considered a front runner;

- Marketing reports to see if our marketing personnel have stressed our performance on similar contracts; and

- Page limitations, special standards, new specifications, and unusual questions for pre-RFPs and RFPs.

Every "no bid" decision contributes to shaping the firm's place in the market by omitting risky, costly, or impractical bids and proposals. Every decision to bid, of course, results in writing a full-scale proposal. Together, the two actions (bid and no bid) should keep the company informed of what is to be written. The crucial part of any technical proposal is the technical approach, because without it, the rest of the document is worthless. Therefore, the bulk of corporate prewriting activities should concentrate on the technical approach that clients are selecting. Again, personally monitoring each client's technical and contracts manager for news of what is happening helps the firm understand why winners win. You should look for the following assets in a client:

- Bold, imaginative technical concepts that present new ideas and offer efficiency and economy;

- Brisk, aggressive management teams that cut the red tape, are cost-conscious, and leave a clean audit trial;

- Similar experience where these new concepts were proven to work, deliver results, and are reasonably free from risk; and

- Personnel who can make it all happen.

In essence, company management must be ready to absorb and replicate successful, useful approaches to services, management, and industrial technology.

5.1 STANDARD INGREDIENTS OF PROPOSALS

We must read and comprehend the clients' solicitations to prepare for answering their RFPs. By studying these solicitations, we see recurring patterns of client interest, and we proceed as follows:

- Sample delivery orders: we analyze three different radar systems and rank them in order.

- We describe how the core staff will interact with client management and how responses to crises will be accomplished.

- We describe how to implement a new military specification for software architecture, such as DOD-STD-2167 and its corresponding specification for software quality evaluation in DOD-STD-2168.

For example, suppose our best client put out three RFPs this year in small, highly technical jobs for which we did not bid. However, the thrust of the questions alarmed us because we admittedly had difficulty answering the new questions. More importantly, the questions provoked internal debate about whether such issues were in our sphere. In fact, we realize that we did not bid because we did not know how to answer the unfamiliar questions. This client is formulating a large, mainstream services contract for which we are well suited, and is scheduling a pre-RFP release in four to five months. We must do the following to get ready:

- Concentrate company thinking on new, industry-wide issues and take a position based on our strengths;

- Hire employees or consultants to close up gaps in our technological knowledge;

- Mobilize marketing to focus on this client;

- Use the Freedom of Information Act to obtain proposal narratives of what is winning in this arena; and

- Be ready to pursue any surprises in the RFP, such as variations on some of the new questions.

In truth, few questions arise from an RFP that we cannot prepare for far in advance. Studying the client's RFPs from the past several years allows us to draw analogies between the client's past performance and the current RFP.

- *Technical concept:* our concept is so new that it has not been proven internally, but the client has confidence in it. The concept departs from the old line of products and services so dramatically that we do not yet know how to market it. It has rough spots that may not be addressed in the RFP.

- *Management team:* our old management method was to bring in one general manager, and hire everyone else locally. The client clearly wants to form a core team. However, only three of seven needed managers are committed to date.

- *Similar projects:* our current projects do not resemble or require the performance level requested by the new RFP.

By reviewing the standard ingredients of older proposals, we know we can answer the RFP. But we will need to keep track of our own internal progress and routinely ask ourselves, "How will we actually perform this contract?" From this self-imposed interrogation, we conceive a new, updated list of items, which the client considers standard ingredients for responsive, winning proposals. We test these ideas against marketing reports to see if we are on target.

5.1.1 A Typical Proposal Outline

We should make a quantitative proposal outline for any RFP worth pursuing. We derive typical clients' standard RFP outline format in Table 5.1. These items consist of a routine study of what 8-10 RFPs would all have in common, that is, ingredients with a high degree of being repeated, hence, expectable.

The RFP usually divides and separates the technical approach. In Table 5.1, the technical approach is discussed in Topic 1 and Appendices A and C. This division of ideas makes it hard for our teams to grasp and unify their own thinking. It is especially bothersome for engineers and technicians who frequently want a "handle on the problem" in the form of one coherent document.

Table 5.1
War Game Scenario for Typical Client's RFP Outline-Estimate

1. Technical Approach to Scope of Work
2. Facilities and Capabilities
3. Quality Provisions
4. Engineering Personnel by Resumes
5. Production Personnel by Resumes
6. Management Approach
7. Similar Experience

Appendix A: Sample Delivery Order
Appendix B: Core Staff Team Implementation
Appendix C: New Hardware and Software Applications to this Scope of Work

---

For every topic and appendix of our sample outline, we fill in as much information as we can, so that we can determine whether we are knowledgeable enough to bid on the proposal. If we indeed understand the RFP, then we decide to bid, and begin pre-writing.

Prewriting tasks should be planned, scheduled, and staffed even if the pre-RFP evaluation has only recently been satisfactorily completed. For each point on the outline, we should formulate technical arguments as to why our team is the best to manage and engineer the work, why our approach is clearly superior, and why our projects are demonstrably successful.

Proposal prewriting preparations include asking ourselves the following questions:

- What actual accomplishments can we show?

- What people do we have whose work shows that we can engineer and manage?

- What proposals have we written for this customer?

- What proposals have we written on similar matters?

- Do we have several *choices* in our client's RFP outline?

In anticipation of the final RFP that the client will release, our staff can attempt to fill in gaps and investigate unknowns.

5.1.2 Typical Proposal Categories

It is advantageous to enter all data for proposals into a secure data base so that we can review, modify, and print out data easily.

Details are an all-important factor in later articulating the ability to perform work in any proposal. A good data base should allow us to crosscheck files so that a search will identify all the new materials on hand. We should break down all proposals into these files:

- *Technical Approach* by discipline (ground water hydrology), skill (nuclear engineering), project name (the ABX Project), and project manager (John Doe's project);

- *Contracts* by client (Army, Navy, Air Force), type (cost-plus-fixed-fee), dollar value, worldwide geographical location, cost categories, and division of labor;

- *Personnel* in terms of managers, engineers, technicians, and production staff; and

© 1996 by Holbrook & Kellogg, Inc.

- ***Quality Assurance*** in terms of quality control, evaluation, testing, and standards and specifications in general.

All data-storing material should be kept in a special library near the office of the proposal managers. Thus, successful prewriting activities require the presence of the following:

- Dozens of recent RFPs and proposals from the customer (or very similar customers);

- Hundreds of resumes from which to choose competent employees;

- Ample references and specifications;

- Marketing crosschecks on client preferences;

- A stack of deliverables documenting our work; and

- Technical ammunition and persuasive arguments about a winning technical approach.

Among individuals who regularly write proposals, there is a fair share of disagreement about what clients actually prefer in proposals, and why clients have diverse and specific preferences. For the most part, business and industrial proposals in the private sector are based on unique criteria that vary markedly from RFP to RFP. However, this cannot be said of so-called government proposals where public funds are expended to fulfill public needs. Most government agencies have developed patterns of select criteria to which they return again and again. These patterns yield a great deal of information to the careful reader. Because so many corporations have or had government contracts, the preferences of the government's representatives have had a pervasive influence.

The following materials were developed through the study and accumulation of over fifty government RFPs. They represent a view of how certain clients *tend* to solicit and judge proposals. These criteria are not all inclusive, nor are they any assurance against change in the agencies' policies or practices. They do represent a clear-cut look at what most proposal evaluators in the government believe to be persuasive criteria for successful proposals. These materials were openly developed through the

Freedom of Information Act (Sunshine Laws) and are usually accessible to any citizen who asks for them. (See Figure 5.1 and Tables 5.2 through 5.6 for a consensus of what federal military RFPs routinely contain.)

Table 5.2
Categories of Essential Information for Proposals from U.S. Industry

Corporation	Volume I	Volume II	Volume III	Volume IV
Martin Marietta Aerospace Denver, CO	*GENERAL* 1. Describe the work 2. Give performance and quality guidelines	*MANAGEMENT* 1. State capabilities 2. Describe facilities 3. Resumes of personnel 4. Transition plan	*COST** 1. Prices per task 2. Back-up information 3. Other costs and expenses, such as subcontracts and rentals	*NOT USED*
Boeing Aerospace Seattle, WA	*GENERAL* 1. Answer 25 brief questions about all business systems	*TECHNICAL* 1. Describe the work 2. Give quality and performance guidelines 3. State similar experience or former work	*MANAGEMENT* 1. Capabilities 2. Manpower 3. Organization chart 4. Resumes	*COST** 1. Prices per task 2. Back-up data 3. Audited fiscal statements
General Dynamics Corvair Division San Diego, CA	*Preface to Volume I* — Federal Representations and Certifications *MANAGEMENT* 1. Company background 2. Management criteria 3. Progress measurement 4. Manpower availability 5. Resumes 6. Configuration 7. Milestones	*TECHNICAL* 1. Tasks to be performed 2. Configuration 3. Description of how the work will be done 4. Summary table of compliance	*COST** 1. Schedule of monthly funding outlays 2. Royalty data, if any 3. Use of government facilities, if offered or available	

* When the federal government funds a program, it owns all the information, drawings, reports, and other data which are discovered, researched, or documented by that program. There are no royalties paid, no patents granted for work related findings, and no funds paid outside the sum given to the receiving institution except those specified under the contract or grant.

\multicolumn{3}{c}{**Table 5.3**}
\multicolumn{3}{c}{**Categories of Essential Information for Proposals — U.S. Army**}
Volume	**Category**	**Criteria**
Volume I	General	1. Describe the work to be done in detail 2. Describe the materials to be used or the services if this is a service contract 3. List tests needed by MIL-SPEC numbers 4. Give complete milestones
Volume II	Techniques	1. Methodology to arrive at completion of work
Volume III	Capability	1. Facilities 2. Availability 3. Adequacy 4. Tooling 5. Configuration
Volume IV	Quality	1. Comply with all MIL-SPEC requirements 2. Prepare all drawings, reports, and other data items per MIL-SPEC requirements
Volume V	Organization, Management	1. Company background 2. Management availability 3. Progress measurement 4. Union representation 5. Guarantee or warranty
Volume VI	Engineering, Production	1. Resumes of staff for project 2. Company overall
Volume VII	Similar Experience	1. Specific company background for the exact type of work in this project 2. Transitions and adjustments to begin this contract
Volume VIII	Cost	1. Standard Form 60s 2. Back-up data 3. Representations and certifications 4. Financial statements (audited)

© 1996 by Holbrook & Kellogg, Inc.

Volume	Category	Criteria
Table 5.4 **Categories of Essential Information for Proposals — U.S. Navy**		
Volume I	General	1. Company background 2. Refer to all special requirements, procedures, and special bills of materials
Volume II	Quality	1. Comply with all needed MIL-SPECs 2. Describe quality assurance and quality control departments and functions
Volume III	Approach, Personnel	1. Management approach 2. Program criteria 3. Company's organization 4. Similar experience 5. Progress measurement 6. Scheduling 7. Availability and adequacy 8. Resumes
Volume IV	Technical Approach	1. Tooling needed *versus* tooling on hand 2. Production plans and industrial engineering 3. Facilities description 4. Capabilities of company overall 5. Availability and adequacy 6. Make and buy policy 7. Identify long lead items
Volume V	Cost	1. Standard Form 60s 2. Back-up data 3. Certifications and representations 4. Financial statements (audited)

© 1996 by Holbrook & Kellogg, Inc.

\multicolumn{3}{c}{**Table 5.5**}
\multicolumn{3}{c}{**Categories of Essential Information for Proposals — U.S. Air Force**}
Volume	**Category**	**Criteria**
Volume I	Discuss Overall Problem	1. Describe work 2. Equipment 3. Production
Volume II	Comply with MIL-SPEC	1. Item Specifications 2. Documents 3. Preparation for delivery specifications
Volume III	Reliable and maintainable equipment	1. More specifications 2. Subcontract controls 3. Repairability
Volume IV	Milestones and schedules	1. Dates given for all goods and services 2. Name all deliverables
Volume V	Quality	1. Quality assurance program 2. Quality control program 3. Special specifications
Volume VI	Company background	1. History 2. Adjustment or transition 3. Similar experience
Volume VII	Personnel	1. Resumes 2. Company personnel overall 3. Personnel approach
Volume VIII	Ancillary	1. Guarantee or warranty 2. All total experience with USAF
Volume IX	Cost	1. Standard Forms 60s 2. Back-up data 3. Certifications and representations 4. Financial statements (audited)

© 1996 by Holbrook & Kellogg, Inc.

Table 5.6 Categories of Essential Information for Proposals — U.S. DOE		
Volume	**Category**	**Criteria**
Volume I	Technical	1. Key personnel by resumes 2. Project management techniques and cost controls 3. Personnel management 4. Comprehensive discussions of DOEs scope of work 5. Corporate (similar) experience 6. Special technical issues (show firm's expertise) 7. Work quality methods
Volume II	Business	1. Organization and management of the corporation 2. Business systems—are they government approved? 3. Financial condition and ratings 4. Corporate experience and past performance 5. Union agreements and ability to cooperate 6. Small or minority business programs 7. Safety and health requirements per OSHA 8. Equal Employment Opportunity compliance
Volume III	Contract	1. Compliance with the DOE proposed contract 2. Suggest any options, terms, or needed clauses
Volume IV	Cost	1. Standard Form 60s 2. Back-up data 3. Representations and certifications 4. Financial statements (audited)

© 1996 by Holbrook & Kellogg, Inc.

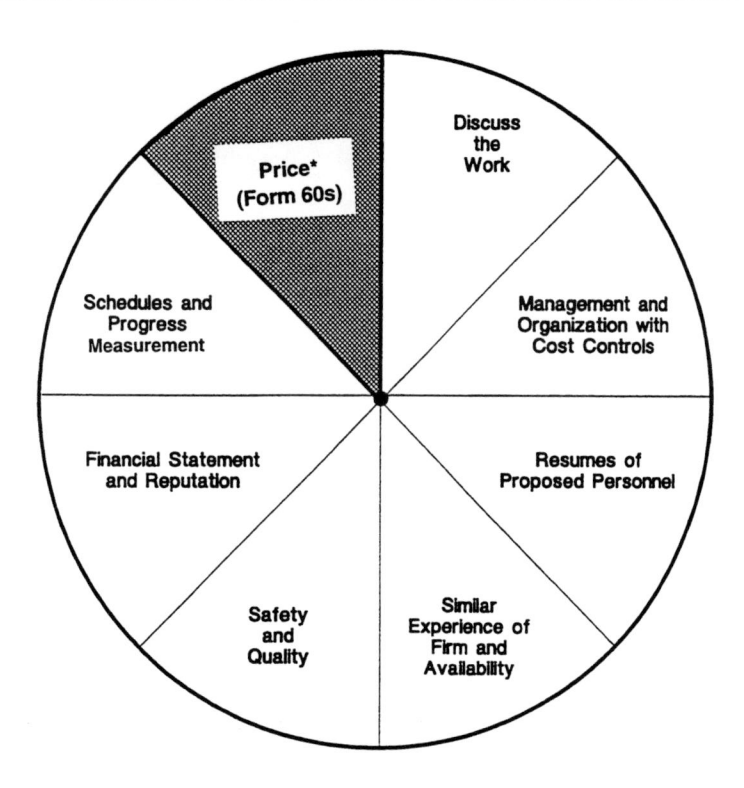

Figure 5.1
Summary of Essential Proposal Criteria

5.1.3 Summary of Ongoing Activities

The customer's primary interests can be encapsulated in this question: "Can this contractor perform the work?" All other factors are secondary. To show clearly that our team can perform, we must do well on the following documents in the prewriting stages:

- Similar experience (performance);

- Resumes (updates);

- "Boilerplate" and "tailored boilerplate;"

- Standards and specifications; and

- Independent research and development feasibility studies and white papers.

Similar Experience. From *similar experience,* a large, successful firm with a 40-year history may have amassed thousands of contracts. A smaller firm has successfully completed hundreds of good contracts. These former contracts are still useful because they point out ways for us to deal with the new RFP. For instance, is the scope of work of the client's new RFP similar to that of an old contract? Did our firm have any nearly identical contracts with this client in the last two to five years? If so, are the staff levels similar? Are the dollar values close? Are the required types of employees similar? Lastly, does the current RFP pose problems similar to those overcome in a previous contract? By reviewing old contracts, we can plot an effective strategy for winning the new contract.

Resumes. Another prewriting activity that our firm must do is to review all *resumes* to find people well qualified for working on the proposal. Not all of our current employees have the expertise to work on the proposal. Therefore, the Personnel Department needs to screen resumes promptly in response to the proposal manager's requests, and identify and update those resumes seriously needed to support the proposal. How does each resume reflect competency in terms of this RFP's scope of work?

Boilerplate. *Boilerplate* and *tailored boilerplate* refer to parts of a proposal such as management plans or quality assurance plans that are usually not written from scratch. They can theoretically be attached to any proposal. However, although boilerplates usually do not vary from proposal to proposal, it is imperative that we scrutinize the plans or materials closely in the prewriting stage to see where and how they might contradict the technical approach. A QA plan for manufacturing, for instance, is wrong for a services contract where no manufacturing is involved. Management plans are often very alike, but may contain faulty steps from a prior contract, which must be deleted.

Standards and Specifications. *Standards* and *specifications* should be a prewriting priority because they take so long to order, receive, catalogue, and distribute. Because standards and specifications change every year or so, a central company library should retain all the documents. The proposal organizations can then use the latest issues dealing with each technical area in the scope of work. It is important in prewriting that we absorb the vocabulary of specific standards and specifications, the accepted flow charts, and the practices and procedures associated with this discipline. Without this familiarity, even a thoroughly experienced engineer can look like a novice, groping for nomenclature and "reinventing the wheel."

IR&D Feasibility Studies and White Papers. *Independent research and development feasibility studies* and *white papers* are another aspect of prewriting preparation, for they flesh out the technical approach, establish expertise, and demonstrate successful applications of new ideas. The proposal team may find these documents from any source—from the company at large, or from consultants, subcontractors, vendors, and business allies. The broader the spectrum of resources, the more complete view the team will have of what is available in the technical marketplace.

5.2 SOURCE SELECTION BOARDS—THE SELECTION PROCESS

The real decision makers in agencies sit on source selection boards (SSB), which annually judge dozens, if not hundreds, of proposals. These senior managers actually choose which proposal will win or not win. Thus, no prewriting activities are complete without knowing who will sit on the SSB of the RFP we have selected. We should ask ourselves these questions about the source selection board:

- Who is the chairman of the source selection board and who manages the committee?

- Who are the other members who vote? Which members do not vote but still advise the board?

- What backup staff will the SSB draw on for expertise?

- Do we know the entire SSB?

- Have we met the entire SSB?

- Have we influenced them all in our favor?

Prior to RFP release, we should visit, lobby, and convince each member of the SSB that our team is the front-runner. Reports, slide shows, and white papers should be compiled and presented before the SSB isolates itself from competitors. We should cover the following points:

- Assessing the educational level, backgrounds, discipline, and technical preferences of SSB persons (A reliability engineer will value different project aspects than a cost engineer or business manager.);

- Determining the SSB chairman's temperament, opinions, and prejudices;

- Praising similar projects in which we showed quality approaches and performance, and used quality personnel; and

- Comparing and contrasting our projects and those of others and planting the win strategy early by showing technical and management superiority, economy, and innovation.

It is important to gauge the responses of the Board members. If one or more members were cool to the approach, were they lobbied by a competitor? Have they had a bad experience with us before? We should ameliorate any problem areas before, during, and after proposal preparation. We should lobby cool SSB members to assure them of no more future problems, and be prepared to structure our proposal to support these promises (Chapter 8). Also, if we discover things in our work that the SSB likes, or things about the competition that the SSB dislikes, we should emphasize these ideas in the proposal, as discussed in Sections 8.1 through 8.4.

We should not rely on one person's opinion of the reactions of the SSB members or SSB official. At least two, and preferably three, competent managers or technical employees should visit or call the SSB. Trip reports consisting of *precise questions* and *attitudes* should be compiled for the proposal team as one of the last prewriting steps.

5.3 PREWRITING AND MARKETING

As we saw in Chapter 2, proposal preparation and marketing affect every part of the cycle that results in a winning proposal. The marketing loop must be closed as the company commits itself to writing a superior technical proposal.

Marketing makes visits to screen the SSB and competition, and to investigate unexplained behavior at the client's agency. Management and marketing confidently inform the client that our firm intends to be the winner. We sift through all data to formulate a win strategy based on our strengths and the weaknesses of others.

Throughout the industry, we tout ourselves as the single most competent firm to perform this work. We offer the client assistance in writing the scope of work for the draft RFP, guidance with types of required deliverable reports, and ideas on similar work from our successful projects. Lastly, we provide the client with consultants, managers, and engineers to demonstrate how dedicated we are to the project. Thus, the marketing loop is closed. Therefore, we can stop preparing ourselves to write the proposal, and press on to the job of actually writing the winning document.

5.4 DEVELOPING AN IN-HOUSE PREPARATION COURSE

An in-house proposal writing course needs the following: a knowledgeable and skilled manager as the instructor, technical experts from inside and outside the company, a well-written text, a good proposal library, and a curriculum that addresses common, recurring in-house communication problems. If the employees can suggest topics and offer solutions to difficulties, then they are more interested in and committed to the program (The steps for an in-house proposal preparation course are condensed and shown in Figure 5.2).

Many engineering and technical firms recognize that excellence in proposal preparation is the focal point of the firm's success. Communication within the company and with customers and potential customers can usually be improved by means of an in-house course in proposal writing. The reasons for conducting such training sessions in house are numerous:

- Sessions are scheduled to the company's convenience;

- Courses are secure and free from competitors' scrutiny;

- Employees' good will increases;

- We can select the material; and

- In-house courses are less expensive than seminars or consultant courses.

> **Productive Courses Consistently Feature Five Key Ingredients**
>
> - Management commitment to a course of certain scope, covering enough material to attract a wide audience in-house, but also encompassing areas where the company perceives that significant improvement is needed. This is a win-win scenario; people get to practice and talk about areas where they excel, but they must also hear about and be appraised of areas in which they have not excelled.
> - A sponsor committed to the course. A manager with sufficient budget and inclination toward the proposal environment should be selected to oversee, plan, manage, and evaluate the course from start to finish. That manager is responsible for a favorable outcome, the assimilation of new ideas, addressing past shortfalls, and other matters germane to proposal improvement.
> - Employee input is essential, a month or more ahead to allow the influx of multiple questions, suggestions and non-managerial concerns.
> - Pertinent materials from past proposals should be incorporated for a sense of realism; also, industry-wide materials, such as other firms' handouts on proposal improvement, should be included so that the course will not be created in a vacuum.
> - There should be an initial course plan, modified by a final plan so that the sponsor and interested parties can watch the course concept grow and develop. There should be a feed-back loop for attendees to filter their comments back to the sponsor and to management.

Figure 5.2
In-House Proposal Preparation Courses

Although many business and professional courses and seminars are available for many disciplines, few of them are as immediate or useful as an in-house course aimed exactly at one of the firm's problems. The in-house course is relatively inexpensive because it saves travel fares, fees to outside organizations, and time out of the work week.

Of course, there is no substitute for a first-rate seminar planned and assembled by a team of persuasive proposal professionals. When nothing short of such a course is required, the company should invest in the travel and expenses. In either case, the company should readily realize how beneficial a course will be.

5.4.1 Scope of the Course

After recognizing that we can improve proposal writing in house, the next step is to establish priorities in the course material. For example, are the departmental technical approaches poorly written and hard to follow? Are business plans negative and difficult to understand? Are the firm's resumes halfhearted and unresponsive to client instruction? What are the obvious problems that need work?

The engineering or technical firm can do a lot to help recognize where the communication barriers are occurring and even diagnose what would satisfactorily solve the problems. After a company has answered questions about its needs, then it can address the issues involving people.

The course now enters the planning stages. A typical planning cycle for such a course might follow these steps:

- Management decision;

- Preparation of scope of the course;

- Selection of sponsor and writing authority;

- Preliminary course plan;

- Lessons learned by the industry;

- Employees' input and contribution;

- Incorporation of pertinent material; and

- Final plan for course.

5.4.2 Finding the Right Sponsor and Authority

Because an in-house course must have leadership, a skilled manager with a strong knowledge of proposals should be responsible for the curriculum. He or she will discuss problems with the other managers; identify problems; make arrangements for the room, word processing support, copying, projectors, and other equipment; and ensure that a course runs smoothly.

Even the best managers, however, have some limitations. It is not surprising that many strong speakers do not write proposals well. Sometimes the company's problems are so unusual and difficult to solve that the company seeks expert help from the outside. (Firms with staff writers and editors have this capacity already and should ask these people to contribute to the course.) Many firms choose communication consultants from the industrial community for this assistance. Others choose a public relations or advertising firm that has some expertise in technical communication.

However, a proposal writing consultant can often be more objective and instructive than, say, an advertising or public relations person. An authority on proposal consulting would make an excellent keynote speaker for overall technical proposal communication background, theory, and practice. Two firms, Judson LaFlash's Government Marketing and Hy Silver and Associates, are well known national proposal firms which can provide such speakers.

5.4.3 Useful Publications

A company's reference library should include the special issue of *IEEE Transactions on Professional Communication* (Developing Proposals, June 1983) and this book. Another fine reference is R.B. Greenly's *How to Win Government Contracts* (Van Nostrand Reinhold, 1983). The classic Jim Beveridge book, *Creating Superior Proposals*, is another excellent reference. The Society for Technical Communication has also published anthologies of excellent essays on proposal preparation. Many mainstream technical writing books have a useful chapter on proposals. The company's old proposals (winners and losers) should also be available to fill out a dedicated proposal library.

5.4.4 The Final Ingredient

A company should thus have five essential ingredients for an in-house proposal writing course:

- A management sponsor;

- An authority in proposal writing;

- Industry-wide models;

- A proposed curriculum; and

- A strong text.

What is missing? Concerned, interested people are probably available who could provide insight and perspective into proposal problems; let them have a voice in the shape of the course and the material to be used. The curriculum should address prevalent problems, which the experts sort out and try to solve. Let the experts diagnose the problems inherent in the materials. In this way, the course not only interests the employees, but also creates a broader, yet specific, basis for information sharing. An additional benefit is that the communication experts will see a spectrum of documents and will be able to analyze pervasive problems. Some of the most common problems in communication occur in these areas:

- Arrogance in writing;

- Lack of logic, and lack of details;

- Difficulty in applying the theory of persuasion to proposal writing;

- Poor grammar; and

- Storyboard approaches that encompass the work of the whole team.

Some people may resist the idea of a course. If these people really write well, they may not need the session. On the other hand, people who are poor communicators do not wish to draw attention to their shortcomings. They should be encouraged to attend, but mandatory attendance is usually counterproductive. Mandatory attendance is required for the entire team of a forthcoming proposal, to practice and train the techniques together.

5.4.5 Timing

Schedule the courses for morning sessions, when everyone is fresh. Limit the session to about an hour for each aspect of the course. Let the material set the limits for the number of sessions; however, fewer than five meetings are usually not very productive. Some sessions may cover the following:

- Improving resumes;

- Locating pertinent similar experience;

- Sizing up the competition; and

- Examining recent winning and losing proposals.

A brief, but professional, flyer should be made and distributed throughout the company, and certificates should be given to the people who complete the entire course. The usual course session has a proposal authority lecture on a topic and then lead a discussion. Managers should dominate the discussion. "Question and answer" periods should be as free and open as possible. If the course is held every year or every quarter, then feedback should be obtained and kept available. Lessons learned from recent victories (and losses) should be a focal point of this feedback.

There should be a prompt and effective improvement in proposals as a consequence of newly polished skills and the solutions to persuasion problems. Employees receive an added benefit; they had an opportunity to contribute to their own in-house course.

5.5 PREWRITING RESULTS

Firms who practice prewriting activities uniformly state that they get a better product, a more uniform product, and a more responsive product from contributors than from putting people into the proposal atmosphere "cold." As urgent, must-win proposal teams are formed to defend or win key targets, the best way to assure compatibility, smoothness, and team spirit is to put the people together in prewriting activities geared to their own unique needs, but moreover to giving them cohesiveness in the face of a tough RFP.

CHAPTER 6

Where RFPs Come From and Why They Must Be Obeyed

We pause for a moment from the craft of writing proposals to a brief summary of where RFPs come from, and why veteran proposal writers consider the RFP their "bible." Every RFP is the work of a committee of government workers putting together agency priorities, policies, legal language, forms of contract, opinions of a diverse nature, and getting it all under one cover.

An average RFP may take one year to complete prior to release; some take two years if a pre-RFP is involved, or if there is public controversy. We break the process into pieces to show the reader that the agency has invested its best efforts in the RFP, and that they intend to be obeyed throughout. At length, we will incorporate how to come to grips with the RFP, incorporating its every element into our own writing plans.

Funding the RFP

Each RFP is "the bible" for its contents. In the federal procurement system, each aspect of the spending of public funds entails:

- Congressional appropriations;

- Subcommittee hearings;

- Government financial management tracking; and

- Performance monitoring of current contractors.

The budgeting and financial process in Congress goes from year to year, providing more money to administer existing programs, and new money for new programs.

Constituency enters in here with the politically motivated, elected representatives being in charge of validating each President's overall federal budget. Each time a major concept emerges—a Seawolf submarine idea, for example—the constituencies of the shipyard states, the submarine electronics industry, and the submarine services contractors come forward to exert positive political pressure to spend money to make jobs. If your firm is a Navy-oriented contractor, with on-going work for the submarine community, then you are part of the process, exerting subtle advocacy for each new program, on its merits. Your firm, as an interested party, "pulses" the tone and speed that appropriations of funds are going through the House and Senate budget committees. Others with competing opinions infuse their thoughts on these funding bills; budgets ebb and flow; time passes, and as soon as money becomes available, it is routed to divisions and offices where it can most pragmatically be used to acquire goods and services. At each turn, marketing personnel must trace progress, watch opponents, and anticipate future steps. It is a full time, perennial job to oversee the Congress.

Major federal acquisitions follow a similar iterative pattern for each agency. At each stage in the federal process, various political battles are fought, won, or lost. The impetus for a major RFP, such as the Seawolf submarine, begins when a military or civilian agency identifies a new, unspecified need to effect some sweeping change in the nation's military or Federal posture. For a year or so, both houses of Congress study, evaluate, and recommend options. Contractors contribute white papers, feasibility studies, and are offered opportunities to testify.

For another year, the studies continue, but now they concentrate on tabulating how much each facet will cost to develop prototypes, do studies, and otherwise evaluate. The agencies create the so-called *project offices* to answer all congressional, constituent, and audit agency questions concerning cost and technical matters.

For argument's sake, let us say that for each successful new concept, four or five other new concepts die in committee, or receive allocations so low they will perish in the feasibility study stage. The careers of appointed or career civil servants ride on these battles for power and influence. Congress follows the struggles keenly for political feedback and backfires.

By the time a project office goes into full swing, there are already enemy camps of people for or against any particular measure allocating public funds. Again, entire careers are created and lost in this highly charged atmosphere. The RFPs start taking shape here from a philosophical stance, with winners exercising the power of a victory over the dissent of losers in the form of preliminary scopes of work.

The advocates of a bill for a new class of missile-carrying submarine, for example, will have a pretty good (if rough) idea of the size, weight, and complexity of the vessel, and even fair (if low) cost projections. At this point, these advocates of undersea power will have done the following:

- Defeated numerous surface vessel claims against subsurface vessels;

- Defeated other services' claims that this vessel is unwise;

- Won congressional and constituent support;

- Won contractor commitment; and

- Won enough of a budget to write a bevy of feasibility studies, each aimed at eliminating risk in the scopes of work.

Preparing the RFP

By the time the activities of prototyping occur, actual contracts are being strategized and mapped out. Again, within each agency, more battles are fought for influence. Philosophies reign supreme, with ideologues of all sorts arguing internally about each turn made. This *formulation stage* shapes pre-RFPs long before they are released as conceptualizations of what the client hopes to get out of an ideal contractor.

In the *justification stage,* factions within the client's firm bare their teeth and claws as tradeoffs occur. Divisions ruthlessly undercut other divisions. Powerful groups replace groups and branches whose motives and power bases have eroded. At the time of release of the actual RFP or pre-RFP, the victorious camp has concrete knowledge of what it wants to see enacted. The vehicles of enactment are:

© 1996 by Holbrook & Kellogg, Inc.

- Architect-engineer contracts;

- Engineering services contracts;

- Design development contracts; and

- Various engineering support services.

These major fields are broken down further into:

- Manufacturing;

- Engineering by discipline;

- Controls—that is, configuration and quality; and

- Project office support services.

All functionaries are supposed to report to and unite in the *project management office* (PMO). The contracts office and contracting officers then issue actual RFPs against longstanding requirements. It is at this point that some inept contractor sends in a poor proposal, its approach stating in effect, "You guys over at the client's office don't know enough about thermal subsea factors; the RFP should say . . . ." Then they propose something completely unamenable to the client's hard-won concept. Of course, this proposal will get little client consideration and is doomed to defeat. Appeals and arguments are fruitless because a large body of contrary data had been generated in support of what the client wanted. *The RFP is cast in concrete and its writers are intent on getting what they asked for.*

Knowing the Client's Needs

To understand the RFP is to know the client well. Once we understand the conflict that spawned the RFPs, we can appreciate how deeply each RFP fits the clients' philosophy, budget, and constraints. We accordingly make our writing plans do the following:

- Respond precisely to each aspect of the RFP;

- Match the "L" and "M" instruction and evaluation criteria fully;

- Assume nothing beyond our verifiable knowledge;

- Avoid reworking the RFP's criteria; and

- Belittle neither the client nor the scope of work.

Let us now link up this RFP data to the writing plan's response matrix as demonstrated in Figure 6.1. Each stem manager must understand the history, meaning, and contents of the RFP, including the:

- RFP requirements;

- Proposal preparation instructions (Federal "M" Section);

- Scope of work or statement of work;

- Contract Data Requirement List (CDRL) or DIDs;

- Specifications;

- Proposal evaluation criteria (Federal "L" Section); and

- Win strategy.

Thus, the writing plan and RFP are linked together to ensure that we get all available points for evaluation, and do so in a professional, nonthreatening manner.

Proposal Manager. The *proposal manager* is a crucial part of the win strategy—the cornerstone that connects the RFP criteria to the writing plan. Between this person and a designated deputy proposal manager, the whole weight of the writing effort will fail. Each subordinate manager and contributor will ultimately look to the proposal manager as the final arbiter, editor, and boss. The deputy is only a helper, whereas the proposal manager must envision, write, edit, manage, and control the quality of each version of the proposal.

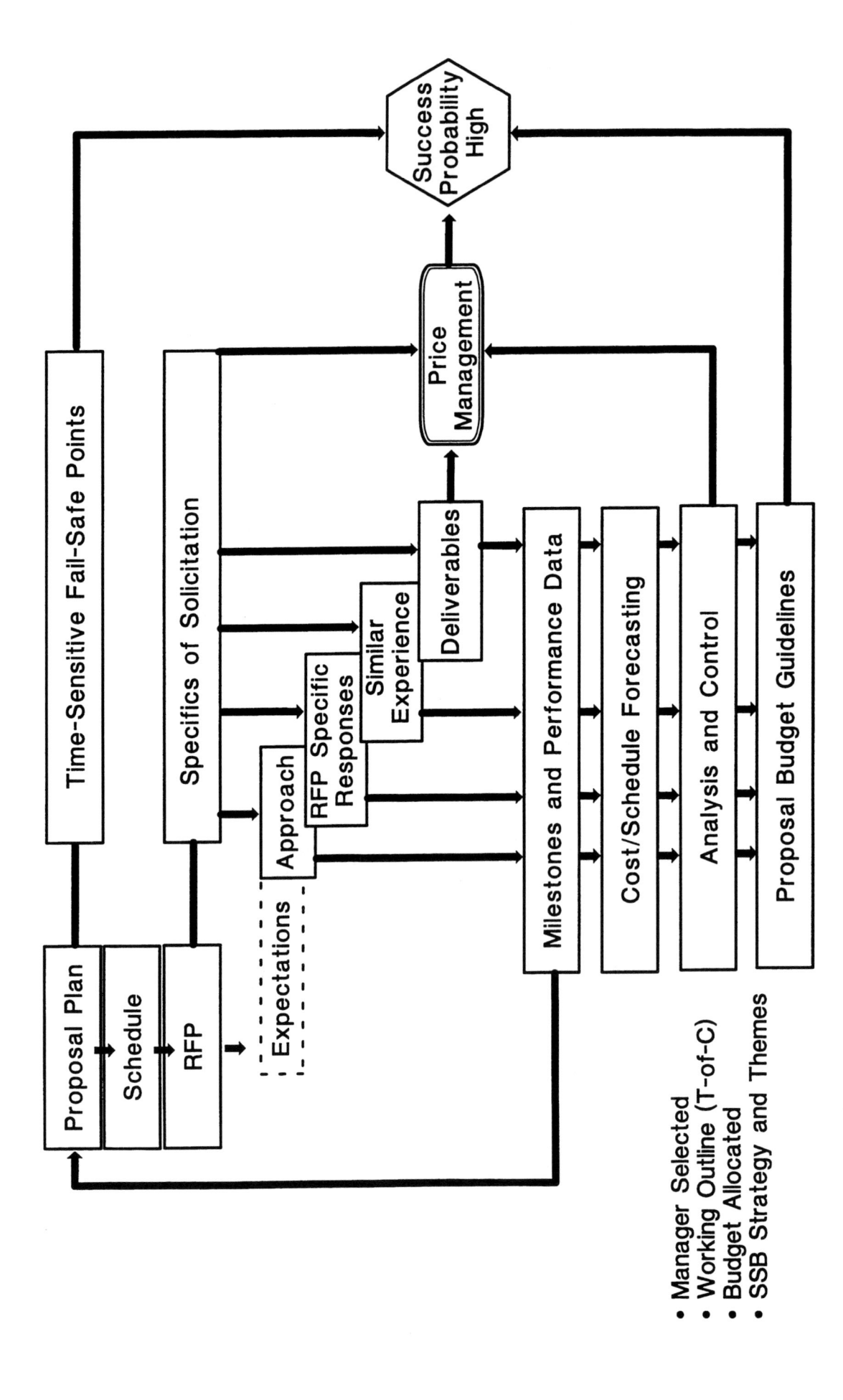

Figure 6.1
Proposal Operations

- Manager Selected
- Working Outline (T-of-C)
- Budget Allocated
- SSB Strategy and Themes

Although this manager's actions are often invisible to the team, they drive the collective effort forward. The mind's eye of the proposal manager envisions the ideal proposal, perfectly tailored, balanced, edited, and ready to win. The vision of the proposal manager must be stamped into the proposal. Because the proposal must read as though written by one person, the proposal manager should be this focal person. By the way, if two or three candidates compete to be proposal manager, who best to serve as Red Team or Blue Team managers than those individuals who were nearly selected to manage the proposal?

Team Members. All *team members* are simply subsets of this manager's concepts. Members are selected according to how well they complement the proposal manager's vision; they help to handle the sheer bulk of tasks, the 45 or so scope of work items, the dozens of potential similar project descriptions, and the hundreds of resumes. These contributors should believe in the manager, the writing plan, and the RFP. A typical, recommended proposal staff structure is shown in Figure 6.2. The team also must be shaped and sized to the manager's perception of the needed work force, and if possible, should be familiar with the manager's work habits, preferences, and proposal methodology.

Quality Control. *Quality control* is the last essential of the winning strategy. The proposal manager personally executes the writing plan, instructing each contributor what to write, and then exercising editorial quality control over the materials. Each subordinate manager similarly applies quality control while dealing with all proposal directives, the plan, the RFP, and the team.

When done smoothly by an accomplished manager, the proposal should flow well, on schedule, and to the satisfaction of upper management. Quarrels, lack of progress, poor quality, and frequent replacement of personnel mean trouble—usually, an unprepared manager who did not generate a plan, an RFP compliance matrix, a winning team, or any written quality.

6.1 RFP MATRIX AND WRITING PLAN

As we have seen earlier, any RFP is the product of intense internal competition at the client's project office or project management office. The RFPs represent internal agreements that the client's staff makes, and then and only then, releases to contractors for bidding. The RFPs reflect political and financial battles won and lost, and their scopes of work are very dear to the people who will evaluate incoming contractor proposals.

Therefore, the RFP dictates how the team should react, how the team should be structured, and what the writing plan should contain. The RFP is "the bible" for the White Team, Blue Team, Red Team, and all other players. It is the rule book to which there is no appeal; any contractor who deviates from the RFP can summarily be dismissed from the competition just for failing to follow instructions.

A corollary to this logic process is that the successful proposer will have grasped that some competent person needs to be assigned to write and manage each point. As shown in Figure 6.1, the RFP will dictate certain key deadlines and contents. Winning proposals then plan and schedule their efforts in a structure geared to getting 100 percent of the evaluation criteria. In Figure 6.1, the proposal manager can show his or her own techniques, add dates, and tailor specific RFPs. The writing plan must then encapsulate each item with checkpoints for quality and timeliness.

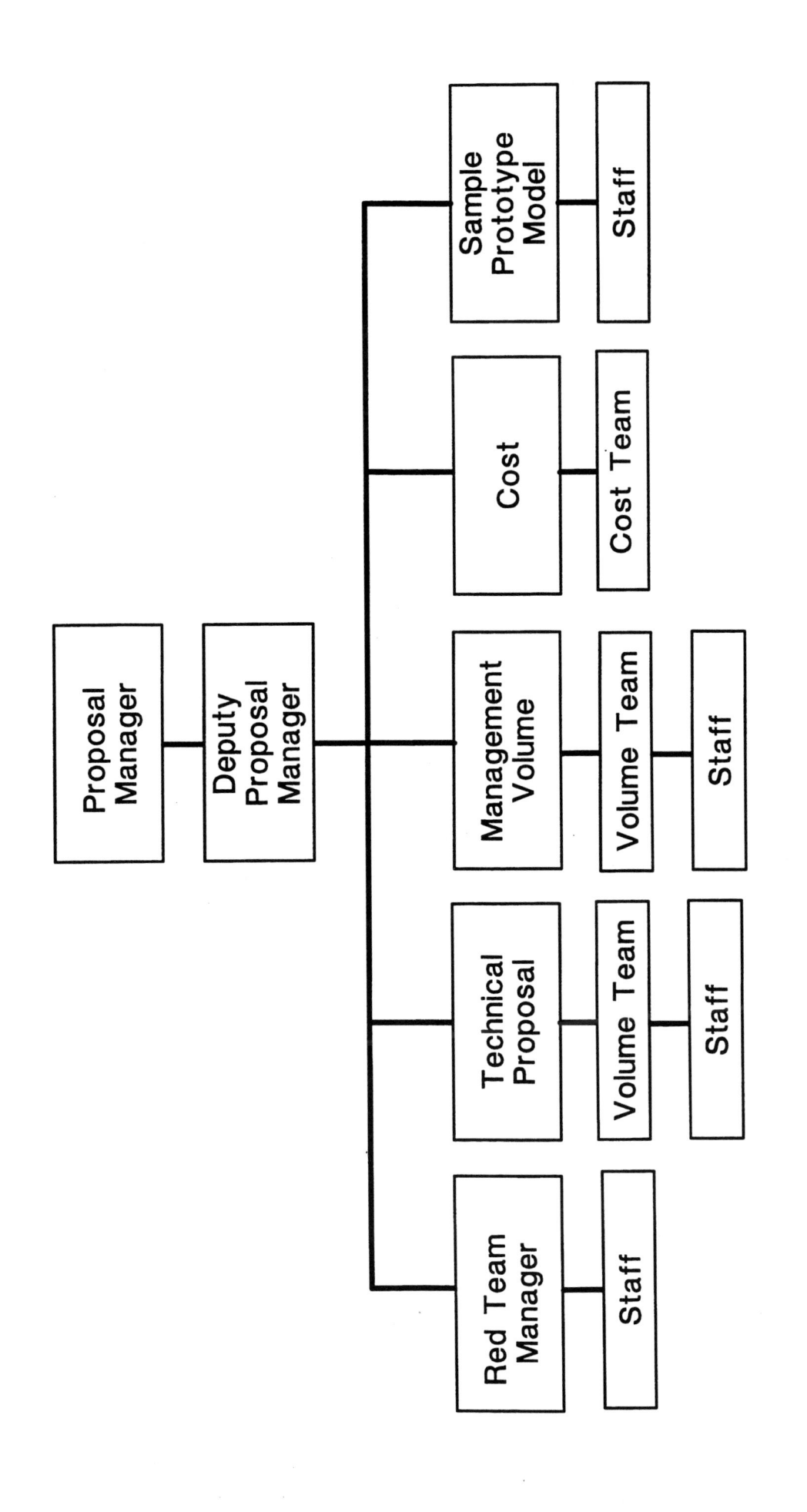

```
                    Proposal
                    Manager
                       │
                    Deputy
                    Proposal
                    Manager
                       │
   ┌───────────┬───────────┼───────────┬───────────┐
   │           │           │           │           │
 Red Team   Technical  Management    Cost      Sample
 Manager    Proposal    Volume                 Prototype
                                               Model
   │           │           │           │           │
 Staff      Volume      Volume      Cost        Staff
            Team        Team        Team
              │           │
            Staff       Staff
```

Figure 6.2
Typical Proposal Staff Structure

Typically, a list such as this captures all the RFP points in a preliminary table of contents based entirely on the RFP, hypothetically assigned, as shown in Table 6.1.

Table 6.1 Preliminary Table of Contents	

1.0   Management Plan
      1.1   Cost Schedule Control
      1.2   Management/Business Plan
	Smith
	Jones
	Brown

2.0   Technical Approach
      2.1   Manufacturing Capability
      2.2   Manufacturing Plan
      2.3   Configuration
      2.4   Quality Plan
	Murray
	Doe
	Brown
	Murray
	White

3.0   Scope of Work Compliance
      3.1   Specific Tasks
      ....
      3.99
	Albert
	James
	Others
	Kilroy

Here the team was dictated by who was best suited to each task by experience, seniority, and priority writing skills. Persons who demonstrate leadership and expertise are placed as volume or section managers because they know the RFP is the heart of the proposal. As the list of assignments is completed, the first table of contents will emerge in a draft form.

However, the proposal manager must be wary of certain situations that hinder a team's optimal performance. First of all, because the RFP calls for certain kinds of people to articulate data precisely, it is unwise to assign members arbitrarily. An engineer put on the cost team may not fit well with the business goals envisioned; a quality control (QC) inspector may be lost working in the design development section. In addition, people loaded with multiple responsibilities are not as able to do well on any one task.

A good manager does not put people on a proposal team just because their department is temporarily underutilized; they may be yanked back to their assignments with little advance notice.

© 1996 by Holbrook & Kellogg, Inc.

Beware of unproductive people who are "on the track;" that is, they are routed from proposal to proposal with management giving little thought of how well they accomplish their assignments.

The RFP is our ultimate proposal priority. The RFP has absolute authority and is the source of:

- Deviations;

- Incompleteness;

- Evaluation;

- Score checking;

- "M" Section and SSB; and

- Working outlines.

There is a chilling tendency among some engineering staff to try to outguess the client's RFP. These mistakes can be avoided if the RFP is used verbatim:

- All emphasis must fall on the RFP's major points;

- Respond exactly to the RFP, item for item;

- Do not rewrite the RFP;

- Do not omit RFP items; and

- Do not make assumptions outside the RFP.

The client usually has checklists to protect itself against proposals that deviate from the RFP.

© 1996 by Holbrook & Kellogg, Inc.

6.2 RFP COMPLIANCE MATRIX CHECKLISTS

In order to further demonstrate what we mean by coverage and matrixing of the RFP, enclosed you will find two actual checklists designed to assist your shop in double-checking and cross-checking RFP responsibilities. The first, Table 6.2 is the Navy's own checklist for bidders; the second, Table 6.3 is Bruce Adkins' checklist, which details in the utmost precision how winning proposals match their RFPs. Together these documents underscore compliance at the optimum point. We then conclude this section by articulating in Section 6.3 how to achieve cohesiveness in the final product.

6.3 IMPROVED PROPOSAL WRITING: UNITY, COHERENCE, AND EMPHASIS

In many phases of modern business, individual success in a corporation is often influenced by how well a person can communicate in proposals. Many technical people find that their jobs require them to be effective in these written presentations and, in fact, there is an increase in written responsibilities as they progress into supervisory and management jobs. However, proposal writing can be a very troublesome matter—some people even avoid the parts of their jobs in which they must write.

The five points discussed below provide a condensed refresher course in improved proposal writing, and they can be used to strengthen and polish this valuable business skill. These points are applicable to the range of documents: technical proposals, business proposals, and feasibility studies (See Figure 6.3 for an overview of how the proposal elements work together.)

Be Persuasive

Your primary goal in proposal writing will be to report on some activity or course of action that you feel is a good, moderate, or bad idea. Be committed to your unbiased opinions of these actions and activity. Do not halfheartedly state your case; you are asking others to think about something as you yourself think about it.

What can you do to persuade evaluators? First, you must know the material thoroughly. Second, you must study the intended receivers of your written work, and you must then aim at that specific audience.

Table 6.2
Sample RFP Checklist

DEPARTMENT OF THE NAVY
CAUTION TO OFFERERS

Prevent costly mistakes by carefully checking your offer *before* submittal to the purchasing office, and *especially check the following points*:

1.	Have you read the Request for Proposals (RFP) carefully? Remember that the RFP provides that the Government can accept your offer without additional discussions with your company (see FAR Clause 52.215-16, Contract Award (APR 1984)).	
2.	Have unit prices and extensions been verified and are they set forth in the proper columns?	
3.	Does your delivery schedule exceed the required date(s) in the RFP?	
4.	Is your offer acceptance time consistent with the requirements of the RFP?	
5.	Have you requested clarification on points not clear to you in writing *prior* to submittal of your offer?	
6.	Have you properly completed the "Representations, Certifications and other Statement of Offerers" of Part IV, Section K of the RFP?	
7.	Are discounts (terms of payment) correctly stated?	
8.	Do you propose to furnish material in accordance with the specifications, drawings or descriptions, as set forth in the Request for Proposals?	
9.	Have you carefully reviewed the packaging and marking requirements?	
10.	Is the person who signed the RFP authorized to do so?	
11.	Are you mailing your offer in sufficient time for it to reach the Purchasing Office prior to the time established for opening of the RFP?	
12.	Does your RFP package indicate the Request for Proposal Number, Date of Opening, Time of Opening, and will it be mailed or delivered to the correct address and building?	
13.	Have you placed sufficient postage on the envelope?	
14.	Have all the amendments to the RFP been properly acknowledged, signed, and returned to the Purchasing Office, and on time?	

© 1996 by Holbrook & Kellogg, Inc.

Table 6.3 Detailed Proposal Evaluation Checklists for Authors and Reviewers[1]			
Step	**OK**	**Revise**	**Comments**
TRANSMITTAL LETTER (GOVERNMENT)			
Correct address			
RFP referenced			
Compliance with all procurement requirements			
Our contacts with telephone numbers			
Alternatives, options, special messages			
TABLE OF CONTENTS (COMPARE WITH TEXT)			
Proper balance and emphasis			
Sufficiently detailed			
Meaningful headings, subheadings			
Figures and tables listed separately			
SUMMARY (OPTIONAL)[2]			
Responsiveness indicated			
Sponsor benefits stressed (concrete)			
Deliverable items listed			
Options or alternatives stated			
Recommended approach (reasons)			
Qualifications (substantiated)			
Superior management shown			
Schedule meets requirements			
Subcontractors, consultants, suppliers			
INTRODUCTION			
Basis for proposal submittal			
Responsiveness indicated			
Special requirements noted; compliance shown			

© 1996 by Holbrook & Kellogg, Inc.

Table 6.3 Detailed Proposal Evaluation Checklists for Authors and Reviewers[1]			
Step	**OK**	**Revise**	**Comments**
Evidence of understanding problem (no RFP imitating)			
Options or alternatives (recommendations)			
Exceptions or deviations presented			
Groundwork laid for later discussions			
Benefits stressed			
Novel ideas highlighted			
Approach planned			
Defects of competitive approaches explained			
Key personnel			
Unique or unusual facilities			
Subcontractor, consultant, supplier arrangements			
Utilization of sponsor's facilities			
Qualifications (substantiated)			
Phases			
References to appendices and proposal sections			
Management interest			
Third-party endorsements			
Conflict of interest compliance			
CROSS-REFERENCE PAGE OBJECTIVES			
Overall objective (sponsor benefits)			
Special subobjectives (concrete and deliverable)			
Listed and numbered (to facilitate discussion)			

© 1996 by Holbrook & Kellogg, Inc.

Table 6.3 Detailed Proposal Evaluation Checklists for Authors and Reviewers[1]			
Step	**OK**	**Revise**	**Comments**
Separate statements relative to each decision point (yes or no wording)			
For phased program, individual phase objectives			
TECHNICAL DISCUSSION — PROBLEM			
Road map (what is going to be read?)			
Background of problem			
Scope			
State of the technology			
Analysis of problem, implications			
Difficulties anticipated (possible solutions)			
Previous related experience (reference to qualifications section)			
Alternatives—advantages or disadvantages			
Options			
Exceptions or deviations (reasons or sponsor benefits)			
High-risk areas (forecasts)			
Trade-off criteria			
Reliability considerations			
Standards and specification considerations			
Regulations considerations			
Interface considerations			
Safety considerations			
Quality control considerations			
Weighing criteria			
Patent considerations			
Policy considerations			

© 1996 by Holbrook & Kellogg, Inc.

Table 6.3 Detailed Proposal Evaluation Checklists for Authors and Reviewers[1]			
Step	**OK**	**Revise**	**Comments**
Capital investment considerations			
"Off the shelf" *versus* special developments			
Standard *versus* unusual procedures			
Field implementation considerations			
Maintenance considerations			
Serviceability considerations			
Training requirements			
Legal considerations			
Production considerations			
Application of technology from another area			
Security or proprietary considerations			
Unrealistic performance requirements (cost)			
Need for sponsor's data, specimens, or expertise			
Cost-effective considerations			
Facilities available—advantages			
Assumptions			
Subcontractor, consultant, supplier relationships			
Testing requirements			
Data acquisition requirements			

© 1996 by Holbrook & Kellogg, Inc.

Table 6.3 Detailed Proposal Evaluation Checklists for Authors and Reviewers[1]			
Step	**OK**	**Revise**	**Comments**
TECHNICAL DISCUSSION — APPROACH			
Selected approach—general description			
Procedures steps (RFP tasks)			
Difficulties—anticipated solutions			
Schedule			
Milestones, decision points			
Critical path			
Review and evaluation			
Analytical and statistical methods			
Samples—number, type			
Quality control procedures			
Facilities			
Apparatus and instrumentation			
Assumptions			
Trade-offs			
Reliability criteria			
Systems considerations			
Interface provisions			
Safety provisions			
High-risk anticipations			
Weighting procedures			
Options			
Alternatives			
Field implementation provisions			
Travel requirements			
Producibility provisions			

© 1996 by Holbrook & Kellogg, Inc.

Table 6.3 Detailed Proposal Evaluation Checklists for Authors and Reviewers[1]			
Step	OK	Revise	Comments
Off the shelf			
Special design			
Performance acceptability criteria			
Subcontractor, consultant, supplier roles			
Utilization of sponsor's contributions			
Maintenance provisions			
Servicing provisions			
Training provisions			
MANAGEMENT			
Management philosophy			
Key management personnel (brief bios, authority, and responsibility)			
Team organization (chart)			
Control plan (PERT, Gantt)			
Progress			
Budget			
Subcontractors, consultants			
Top management interest, access to			
Management experience			
Benefits of interdisciplinary staff			
Procurement practices			

© 1996 by Holbrook & Kellogg, Inc.

Table 6.3 Detailed Proposal Evaluation Checklists for Authors and Reviewers[1]			
Step	OK	Revise	Comments
CAPABILITIES			
Our general description			
Up to date			
Germane points only			
Personnel			
Appropriate—connection made			
Germane experience stressed (technical, management)			
Pertinent publications			
Pertinent societies			
Pertinent awards, citations			
Length appropriate for role			
Consistent format			
Facilities			
Germane only (specific contracts with details			
Organization into appropriate fields of experience			
Appropriate aspects highlighted			
Up to date (current *versus* closed)			
References to proposed staff involvement			
Broad implications (insights)			
Experience			
Germane only (specific contracts with details)			
Organization into appropriate fields of experience			
Appropriate aspects highlighted			

© 1996 by Holbrook & Kellogg, Inc.

Table 6.3 Detailed Proposal Evaluation Checklists for Authors and Reviewers[1]			
Step	**OK**	**Revise**	**Comments**
Up to date (current *versus* closed)			
References to proposed staff involvement			
Broad implications (insights)			
COSTS			
Method of calculation			
Appropriate level of expertise			
Appropriate level of supervision			
Optimal use of subcontractors			
Travel: off-site residence justified			
Options and alternative cost implications			
Purchase decision implications			
Use of sponsor's contributions			
Completion, delivery schedule			
FORMAT			
White space adequate			
Figure and table text references			
Dividers for major sections			
Errors (typographical)			
Meaningful table headings, subheadings			
Meaningful figure captions			
Meaningful text headings, subheadings			
Originality			
Paragraphing adequate			
RFP requirements followed			
Major portions self-sufficient, separable			
LANGUAGE AND STYLE			

© 1996 by Holbrook & Kellogg, Inc.

Table 6.3 Detailed Proposal Evaluation Checklists for Authors and Reviewers[1]			
Step	**OK**	**Revise**	**Comments**
Sentences—simple, clear, concise			
Vocabulary (grammatically correct)			
Proper technical level for readers			
Positive, enthusiastic, confident			
Sponsor-oriented ("you" viewpoint)			
Consistent terminology			
Section, subsection introduction adequate			
Section, subsection transitions adequate			
Section, subsection conclusions adequate			

[1] Reviewers to be supplied with complete RFP and clean draft copy.
[2] If proposal does not have summary, then items may be included in introduction.
Source: Bruce Adkins of Adkins and Associates, Arlington, VA. Reprinted with permission.

Generally, a presentation in technology is a statement of factual data, given in a logical framework. This framework always has a definite beginning, a middle, and an end. The subject is divided into units so that it is easy to understand—not only to accommodate less educated members of the audience, but also to allow hurried readers to grasp the message quickly.

One good device for convincing the readers of your presentation is to *choose concrete examples and words to express yourself*. Such choices demonstrate you are not a "pie in the sky" thinker, and that you are not putting the reader through an academic exercise.

Another excellent rule of proposal writing is *use simple, direct, and straightforward presentations*; do not be coy or tricky with the readers, or they will be tempted to think you underestimate their knowledge or that you plan to mislead them.

Then, *proceed from the familiar to the unfamiliar*. Do not expect readers to be psychic or to have a crystal ball handy. They need certain background, history, and references to records or other technical documents. You need to lead them through the thought and logic process that persuaded you to think as you do—by omitting one of the intervening steps, you effectively break the unity, coherence, and emphasis at which you should be aiming.

Be Logical

Successful proposal writing is always well thought out and follows a logical outline. Approach the subject on a problem-to-solution basis; state the problem, analyze the problem, propose the solution. Be sure to make your point of view clear at the beginning, then hold yourself to a uniform, consistent tone from beginning to end. When the problem is so large and so complex that it appears unmanageable, break it up into logical sections and subsections that can be handled. With extremely complicated subjects, identify and number the sections so that discussions pertaining to them can be tracked.

Check yourself from time to time by looking at your outline to see if you are on the course you planned. Do not force yourself to write the introduction first and the end last; write what seems logical to you, then go back and polish the parts so that they will smoothly fit in your organizing outline. Use illustrations where the text seems to need them, and take plenty of time to work them in. An overview illustration, flow charts, and other graphics will underscore the unity of your writing.

Control the Presentation

In first-class proposal writing, there is no place for random classification of materials, nor for sloppy arrangements or words. When you are writing a longer document, like a technical proposal, there is often a short statement describing the report. Such a brief set of words is usually called an executive summary. Its only purpose is to give a brief, thorough overview of the entire paper. (If this were a memo, the synopsis or abstract would simply be called "Subject.") The summary draws from all the component parts of the work, explaining the whole and complete significance of the entire work. It is a miniature of the proposal itself.

This sharp initial focusing of your material is, indeed, another rule of superior proposal writing, because each paragraph of the material following the abstract or synopsis should be controlled by integrating statements called thesis sentences. Each thesis emphasizes what is of significance in that paragraph in a concise manner; the paragraphs then amplify and expand and give examples of their own respective theses. Both synopses and theses should be brief and to the point.

The theses will aid you in coordinating the important material and subordinating the unimportant. Transitions are then used between major blocks of information, where changes in content are being made. Summaries occur at points where you choose to reiterate emphatic points, and where you want to remind the readers that you are moving from one major block to another. Pass the entire work through several rewrites until it is smooth and controlled. Rewrite the awkward and confusing portions until you are satisfied that each reasonably fits its respective synopsis and thesis. A final word on the control of presentations—never hurry or rush yourself. Plan ahead.

Review the Presentation

Be your own editor on the early drafts. Have you been true to your intents? Were you orderly and organized in the arrangement of the pieces of information? Have you developed and built the ideas? Did you avoid random presentations? When you are satisfied that the paper makes sense, put your eye to the technical side, to the uses of language.

Simple Words. The first rule is to choose simple, Anglo-Saxon words over longer, polysyllabic Latin-Greek words. This stops you from unnecessarily complicating the text. *Measure* things, do not *quantify* them. Did you overqualify the text? Too cautious? Were you speaking "legalese," trying to impress the reader with your "heretofores" and "thereinafters?"

Remember the proposal writer's axiom—write to express, not to impress. Then, make word choices that have no unwanted connotations. Are there angry or hostile words in the text? Are there words aimed at evoking sympathy from the reader? Did you use "shop talk," jargon, or malapropisms that could be ridiculous in their context? These are not proper for professional business documents, and all should be replaced with no-nonsense word choices.

Active Voice. Prefer the active voice over the passive voice. "I saw the profits" is a stronger construction, for example, than "The profits were seen by me." Basically, you should choose strong action verbs over weak "to be" verbs. Next, avoid superfluous, overzealous sentences and do not get involved in outright "padding" of the paper. Are you one of those people who believe "more is better?" Do you really expect the reader to wade through an endless sea of words? Any reader gets impatient with excesses, and wordiness can be the curse of proposal writing. Rewrite for conciseness and precision. Always remember that you are writing for a busy reader.

Momentum and Pace. Next, consider the question of momentum and pace—these are often overlooked by unskillful proposal writers. Momentum simply means that you are stating a point, explaining it as thoroughly as necessary, making transitions out of a point, introducing new points, and making definite progress to a conclusion. Pace is the rate and depth at which new evidence and strategic restatements are being made—you can overwhelm your readers with lightning-fast movements and lack of detail, or you can bore them with too much detail. Good writers use a medium speed that satisfies the needs of the audience.

Grammar. Use grammar as a business skill. Did you leave any dangling modifiers? Did you mix singular and plural nouns and verbs? Are the reference pronouns correct? How about spelling? Without this close, professional scrutiny of your words, your work may appear juvenile to some readers.

Proprietary Information. Now for a necessary step in certain sensitive jobs. Did you touch on or describe something that is covered by a proprietary restriction or by an industrial or military security classification? Did you meet the proprietary or security requirements for dissemination? If not, reconsider your position and rewrite accordingly.

At the end, stop being your own editor; turn the paper over to someone else who is interested and qualified to make comments. The other person who steps into the reader's role now has a chance to give you a fresh perspective. Listen. Be responsive. When you have incorporated the other reader's thoughts, you are close to completing the job. If time permits, additional readers (the Red Team) can provide extra comments and help.

Make a Final Evaluation and Revise Accordingly

At the end of any significant proposal, many people are so wrapped up in the work that they fail to make a last rewriting effort—which can mean the difference between not so good and good proposals. Go back to the criterion of logical presentation: if you were getting this paper from someone else, what would you think of it? Is it truly persuasive? Did you conceal or disguise bad news or setbacks? Did you say what you meant to say in the way you meant to say it? Did you avoid the issue or omit data? Were you pretentious? In rewriting to make the improvements that your reader suggested to you, did you fix one area at the expense of another? Are the unity, coherence, and emphasis still there?

There are no secrets to good, improved proposal writing. In fact, by using the brief points made in this article, most people can refresh their abilities in just a few minutes. The trick, though, is not to let your proposal writing skills get too rusty between refreshers.

L and M Coordinates

We can tell where unity, coherence, and emphasis will be important in Federal contracts by assessing the RFP to see the following:

- Where the "L" Section tells us the government's instructions, and how we should prepare our proposals; and

- Where the "M" Section (evaluation criteria) tells us how the source selection board will evaluate our proposals.

The scope of work dictates three strong quality assurance tasks, which are correlated to another RFP criterion for a quality program, with secondary emphasis on QA deliverables in quantity. These indicators in "L" and "M" tell us the client's value and where to concentrate our emphasis. Figure 6.3 shows how L & M factors "layer" to highlight where the client has emphasized one area over another.

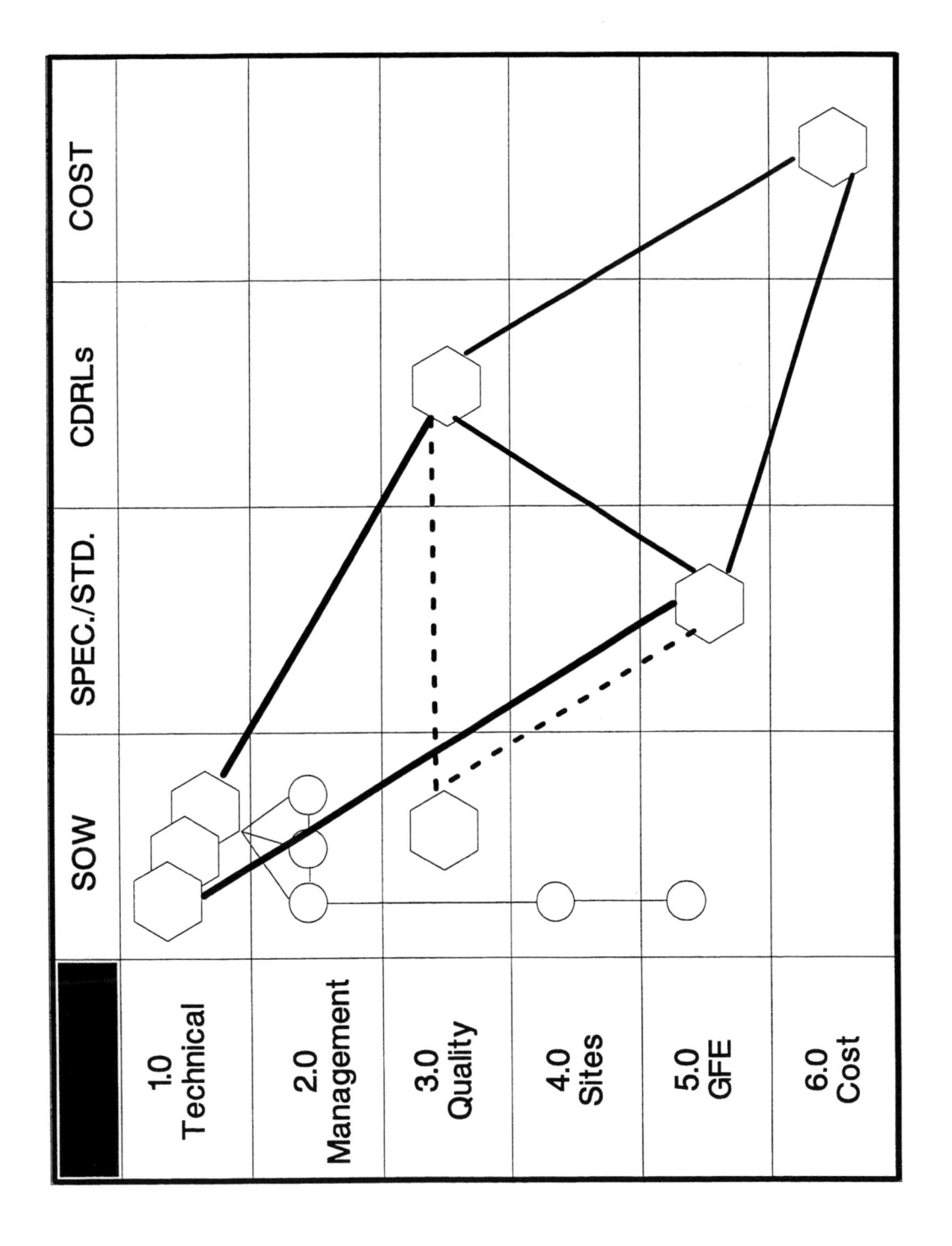

	SOW	SPEC./STD.	CDRLs	COST
1.0 Technical				
2.0 Management				
3.0 Quality				
4.0 Sites				
5.0 GFE				
6.0 Cost				

Figure 6.3
L and M Coordinates—Emphasis and Economy

6.4 WRITING PAST PERFORMANCE & SIMILAR EXPERIENCE: THE PROACTIVE ARTICULATION OF CORPORATE CREDENTIALS

We have itemized above how to write persuasively in the technical and management areas. Now we turn to ensuring exactness in presenting company qualifications, credentials, and abilities as expressed in the Evaluation Criteria of RFPs as Past Performance and Similar Experience. Here, we refer to Past Performance and Similar Experience interchangeably as the parts of a proposal that attest to the firm's established skills, proven in contract performance. Standard FAR practice requires every RFP to contain an evaluatable section called either Past Performance or Similar Experience.

This section provides a means for the government's Source Selection Official to judge potential contractors' on-the-job performance of their work. Figure 6.4 shows a format an RFP mandates on contractors. On the surface, filling out the form looks like a cut-and-dried matter, devoid of articulation, and with little latitude from the contractor's perspective. Nothing could be further from the truth. In fact, today it may be that Past Performance is the most important part of the RFP, with a purposefully designed set of Past Performance and Similar Experience cross-references ingrained in the criteria for technical approach, management concept, and resumes throughout the evaluation checklists that federal RFPs list in Section M.

Similar experience is disguised elsewhere in the RFP's evaluation criteria so that, in effect, the entire evaluation process can be fairly described as a search for the proposer whose experience is ideal to the particular statement of work. The government's evaluators will privately use specially designed checklists to provide their true opinions of a firm's ability to perform on time, on budget. The checklists seek precise levels of detail from each proposal, using YES/NO logic to isolate peripheral proposals from mainstream proposals. By reducing the proposer's experience to YES/NO logic, the evaluators get direct answers to their questions. If the RFP is from the Marine Corps, does this proposal cite Marine Corps contractual experience as a prime or subcontractor? The answer will be either YES or NO, never MAYBE. This inquiry defines whether (or not) a firm is perceived to have experience working with a client; those without proven contract experience can expect to be rated lower since there is an implicit delay and learning curve in educating them to the client atmosphere.

© 1996 by Holbrook & Kellogg, Inc.

Subsequent checklist questions refine the proposing firm's actual knowledge base, and help shape the evaluators' opinions of how easy or difficult this firm would be to work with. Unclear, ambivalent or erroneous past performance or similar experience information and statements from contractors are usually held against them—the evaluators do not make any assumptions or presumptions beyond what is in the written narrative of the proposal and what is in their own checklists to evaluate.

It has been said that the genuine reason for a Past Performance or Similar Experience Section of an RFP is to give an agency's contracting staff the opportunity to compare and contrast agency and interagency performance data about existing and potential contractors unimpeded from outsider scrutiny—that is, an atmosphere that is intrinsically subjective. Past Performance and Similar Experience are to corporations what reference checks are to individuals. They are listed in the belief that every reference (or contract citation) will be favorable in its entirety. Sometimes such an assumption is valid, sometimes not, but the information passed in both personal reference checks and in the Similar Experience conversations will feature distinctly personal and private information, as well as positive and negative types of questions. No reference check would be complete unless the question was asked as to the candidate's ability to get along with clients, punctuality, and quality of work.

Firms can also expect that government contracts personnel will probe deeply into their firm's good and bad performance features. Because individual human beings fill out the government's project evaluation sheets, contractors can anticipate what will be said. (On most cost-plus-fixed fee or cost-plus-award fee contracts, the client staff provides a detailed sets of contractor evaluation reports, technical directives, modifications to the contract, and fee recommendations that reveal their level of satisfaction.) This data must be included and accommodated in Past Performance and can be cited as root causes for nonselection.

The Key Is Currency. The key to writing good Past Performance and Similar Experience sections is currency. All too many firms treat Past Performance as canned boilerplate, dropped in one proposal after another without concern for the accuracy of the data or even for the fact that addresses, phone numbers and contracting officers have changed. It must be noted that in 1995 many federal agencies began printing a notice in their RFPs that they are under no obligation to correct errors found in a firm's experience section, and that they will not go on a telephone calling expedition to track down references whose phone numbers are incorrect. Instead they will inform bidders that a wrong number will equal no score for that contract citation.

Consider the fate of a proposer who was allowed to reference only five contracts, and two of the five citations contained phone numbers, names, and addresses of persons who had been transferred, retired, expired, or moved. This proposal could then expect only 60 percent (3/5) of the experience credit that more careful contractors would receive. Such an omission could be the fatal tabulation of points that puts a firm out of the Competitive Range. In summation, evaluators do not have any obligation to do the contractors' work for them and will score erroneous data and omissions against the contractor's ability to understand the work.

The Citations Are Limited by the RFP

The RFP usually mandates that only five citations can be given and that those must be within the last five years. That means the evaluator contracts older than five years are invalid for evaluation; moreover, if they asked for five contractual examples, don't try to give them more. This gives the contractor an opportunity to do some "cherry picking," to exercise selectivity in presenting ongoing and recent contracts. Contractors themselves can then select their own credentials to articulate, and the client will not per se dictate what contractors will report on.

There are exceptions. The RFP will in all likelihood direct bidders to make their choices "relevant." This is a word with broad implications. For example, if your firm is the single incumbent or one of a consortia of perceived incumbents and you are pondering whether to report the current contract, its omission could be interpreted as fear of evaluation. Omitting a well-known contract to include obscure or small contract is, again, a choice the contractor has to live with, especially at orals when the contracting officer asks how you selected your five citations. The unspoken expectation is that prudent contractors will showcase their contracts and avoid discussing troublesome ones.

However, it is the government evaluator's job to find those projects where the contractor is operating under heavy pressure to perform, where unexpected events forced the contractor to improvise, and where problem personalities have created less than ideal situations. These so-called real-world inputs allow the evaluators to gauge performance on a new contract they suspect will not be easy to perform. The RFP

L and M sections prescribe, however, what the client demands the contractor has to provide on penalty of being found nonresponsive.

It is essential to remember that each proposal has to prove that the contracts included are indeed relevant, driving home to the evaluators that an ideal or near-perfect match has been made. The extremely positive, self-serving rhetoric of sales brochures and annual reports will be viewed with skepticism and evaluated as unsubstantiated if the same kind of narrative is used in place of a technically accurate past performance citation.

What Are They Looking for? Who's the Lowest Risk? As we have said throughout the book, it is up to the proposal writing team to design, develop, and deliver clear narrative to the client, narrative that can be clearly evaluated. In most commercial firms and all federal agencies, selection is made by closely checking what the contractor asserts is nearly risk-free job performance—that is, the company's claims of excellence in performance versus what the civil servants assigned to the project have to report. The two are rarely identical. Subjectivity, personal opinions, disputes, and honest errors in memory all affect the process. Some firms believe they have lost contracts because they had enemies at the client's office; some firms *know* they lost because they made enemies at the client's shop. Persons who distrust a firm traditionally score that firm a high or medium risk; persons who trust a firm grant it low-risk evaluations. Risk measurement in itself is hard to gauge, as we show below.

In contracting, there is an expression about contractors, "There is the Devil we know, and the Devil we don't know." In layman's terms, the devil in this instance is the selected contractor, the firm that has the job now, the devil that the client now deals with routinely. Sometimes that devil is late, over budget, rude, sloppy, and arrogant. Sometimes that devil makes the agency look heroic, saves them face, takes the blame for institutional shortfalls, and makes amends every quarter and every year. When the contracts personnel evaluate proposals, they are always evaluating the risk a new contractor brings along: Will a new firm be later, more expensive in the long run, more stubborn, less responsive, and harder to work with? Would a new devil be any better at all, given the risks, costs, delays, and unknowns of a changeover period?

Risk can be contractually defined as anything that makes the agency look bad, and selecting a firm that fails to meet its statement of work and delivery schedule are chief among those failures that draw adverse attention to the agency. Risk can be defined as simply losing experience—that is, the agency consciously allowing project memory to depart, causing an expansive, expensive phase-in, the loss of knowledgeable employees, the repetition of procedural development, and overall schedule slippage, just the kinds of items that get agencies on Congress's Poor Performers' List and make them targets for a General Accounting Office (GAO) audit for fraud, waste, and abuse. This scenario is the nightmare of every contracting shop and a yardstick for risk measurement. So another definition for risk is uncontrolled, unpredictable change in the contract environment, that is to say change that occurs without the foreknowledge and approval of the contracts shop.

In an effort to eliminate risk, then, evaluators are seeking safe, predictable, reliable firms whose references will bear scrutiny and whose continuous performance has been unmarred by major flaws, errors, or omissions, firms that are cooperative, benign, and easy to work with both managerially and technically. By measuring risk in these ways, agencies can sort out their devils, always selecting the least troublesome ones.

Using the YES/NO Checklist

There is a very large difference between saying (a) "My company needs the revenue from this contract," and (b) "My company is exceptionally well qualified to perform this contract." The clients want to disqualify firms whom they perceive to be saying (a) so that serious evaluation can proceed with firms perceived to be saying (b). The YES/NO checklist logic mentioned above is a useful tool to compartmentalize, isolate, and identify exact pieces of knowledge about just this one contract and its peculiarities. Typical checklist questions follow.

> IS BIDDER'S EXPERIENCE WITH THIS AGENCY?
> IF NO, HOW SIMILAR?
> HOW DISSIMILAR?
>
> IS CONTRACT CITED THE CONTRACT UNDER CONSIDERATION?
> IF NO, HOW RELEVANT?
> WHAT IS MAGNITUDE?
> WHAT IS LEVEL OF EFFORT?
> WHAT IS SIMILARITY IN PERIOD OF PERFORMANCE?
>
> DESCRIBE MAJOR SIMILARITIES BETWEEN CITATION AND THIS CONTRACT.
> DESCRIBE MAJOR DIFFERENCES BETWEEN CITATION AND THIS CONTRACT.

© 1996 by Holbrook & Kellogg, Inc.

IS THE SUBJECT CONTRACT ONE OF A FAMILY OF CONTRACTS?
IS SUBJECT CONTRACT A STANDALONE, GENUINELY UNIQUE PROGRAM?

IS BIDDER ONE OF A FAMILY OF BIDDERS, INCUMBENTS?
HAS BIDDER EVER WON AN AGENCY CONTRACT?

COMPARE BIDDER'S EVALUATIONS TO:
    INCUMBENT(S)
    OTHER CONTRACTORS ON AGENCY'S SIMILAR WORK
    OTHERS IN INDUSTRY

On the federal side, these inquiries are made at great depth, with some contracting personnel amplifying these questions with subset questions of their own made by:

- face-to-face interviews at their own or the other agency's offices;

- assessments of mail/fax questionnaires from other civil servants; and

- telephone interviews based on questionnaires, amplified by additional inquiries.

Evaluators tend to include their own personal findings amid the evaluation factors, accounting for each piece of data under an apparently logical criterion. A wise contractor will be aware that some of the source selection staff will use the kinds of probing inquiries of an investigative reporter, delving deep into how and when opinions were formed. It is safe to say that most commercial and federal evaluations of their contractors are inflated, face-saving, and diplomatic. In the 1980s and 1990s, for example, many DOE contractors routinely received over 90 percent of award fees year after year, yet the same firms were just as routinely rejected, rebuffed, or downscaled when their contracts came up for renewal. Glowing praise in the contractor's Annual or Quarterly Performance Report does not always equate to the agency wanting the same contractor on the next period of performance. Some contracting officers will not stop questioning other offices until they uncover at least some unflattering information about a new and unknown bidder. Such inquiries are leveled at all bidders to even the playing field. Competent incumbents, as we said in Chapter 3, would still have an advantage.

Preparing Past Performance Sections

The contractor should visit and review performance with each referenced contract, just as an individual would clear a personal reference with another individual. Any gap or misunderstanding should be dealt with long before an evaluator double-checks the assertions in the proposal. On rebids of incumbent contracts, each responsible manager should be sure that his/her equivalent on the government side has no hidden agenda, is not confrontational, and will give a good reference. Management should act to remedy sore spots before evaluators turn up areas of disagreement. Attention must be paid to assuring that kudos, positive performance, awards, TQM achievements, agency and civic recognition get mentioned in the citation.

The narrative of each contract citation must be accomplishments-driven: What did you do? When did you do it? What high points and special achievement did you meet on or ahead of schedule? These factors make it easy for an evaluator to believe claims of excellence. If there is no page limit for the citations, try to compact the history of the period of achievements into two to three well-written, crisp pages of technical narrative. If awards, letters of commendation, and official praise are in the record, use those documents as illustrations within the section, proving excellence in the client's own words.

Past Performance Implications Are Throughout the RFP

A firm's past performance and similar experience are not located and evaluatable solely in the proposal section entitled Past Performance. In fact, RFPs spread experience factors throughout the Statement of Work, the Deliverables, and the Instructions and Evaluation Criteria. The infusion of proven experience as a selection factor uniformly throughout the RFP is sensible, logical, and evaluatable. If the team doing the proposal evaluations is to find that one ideal firm for the new contract, they have to eliminate others. Risk, as discussed earlier, can be defined as a lack of experience, lack of expertise, unfamiliarity, and uncertainties. These add up to a proposal that may make a short list of three to five fairly well-qualified bidders, but not score highly enough to be the winner. Where exactly are elements of Past Performance located in the RFP, and what can be done to deal with them? The answers are below.

Similar Experience/Past Performance in Technical Approaches. As the bidder describes how a technical Scope of Work has been planned, managed, and engineered to meet client specifications, it is absolutely necessary to state the ways the bidder will use and abide by the agency's standards and specifications. A new or inexperienced bidder will misstate, misuse or omit standard practices, methodologies, and procedural steps. When scored by the evaluators, these errors and omissions accrue to a demonstrated lack of skill, testifying to a lack of qualified, technical personnel. In brief, the evaluator can now present such a bidder as unqualified and have the hard evidence to sustain protests to the contrary.

Some RFPs ask offerors to provide a Sample Task Order, showing how they would put together a work sample if they had the contract in-house. Again, how similarly the contractor can show on-the-job experience will underscore how well received and evaluated this sample will be. Errors and omissions here will multiply negative findings that evaluators may have located in the technical approach.

Similar Experience/Past Performance in Resumes of Key Personnel. In a free enterprise economy, any contractor can nominate any person to head any project. The only restrictions are prudence, decorum, experience, and reputation, with the emphasis on prudence. The most important resume in any proposal is that of the program manager, the person who will interface with the client almost continuously, who will convey personally an image of the firm, and who more than any other person shapes the direction and performance of the program. Therefore, the program manager's resume should not be boilerplate. It should be tailored to the RFP and emphasize specifics from the Scope of Work. In brief, a summary of common mistakes contractors make on program managers' resumes are proposing:

- a stranger to the client;

- a sales and marketing person as a program manager instead of a technical person;

- an exceedingly technical person instead of a manager for program manager;

- an ex-Army officer for a contract where it is known and expected that the client recognizes that only Naval officers will have the desired characteristics, or a variation thereof;

- a weak or poorly qualified deputy because the former program manager quit or was fired; or

- a fired program manager from a larger programs who is likely to bring their temperament to other projects.

Taken in the same order listed above, an evaluator will score and judge such program managers negatively and as defective or inefficient because (a) the stranger or outsider brings no knowledge to the nuances of the job and doesn't know what is risky or important; (b) the marketing person will market, not manage; (c) an overzealous engineer will design, not manage; (d) if a military service wants one of their own, it is imprudent and risky to disagree with them; (e) a deputy needs to be proven before being thrown in to sink or swim—the risk of drowning is too great; (f) managers with a personal record of dismissal have an industry reputation of running large risks with the client, including personal confrontations. Every skilled evaluator will scrutinize every program manager for these traits and shortcomings.

After selecting a candidate who is free from these defects, that resume should track to the identical Scope of Work or a similar project's Scope of Work, giving for each program listed chronological achievements, exact contractual details, numbers of managers supervised, total staff, budgets, and benefits derived. The government will conduct reference checks on this person contacting past clients, both on the technical and contractual/management sides, to assess his/her personal qualifications and reputation. Again, the savvy evaluator will seek to determine whether this candidate really

- has or ever had technical knowledge of all aspects of the program;

- has educational credentials suited to this management level;

- is reliable;

- is reputable;

- has had cost overruns;

- has had show cause or default or terminations for convenience on contracts, amendments, or task orders;

- has been visited by the IG, DCAS, DCAA, and GAO for reasons others than ordinary;

- is personally solvent and respected in the community.

© 1996 by Holbrook & Kellogg, Inc.

The evaluators will also look for the nature of the deliverables this person has had technical cognizance over—if a project will require completed software programs in a difficult language like Ada++, how many has the candidate finished? Does this manager delegate all the technical tasks to others, i.e., is not practicing at a competent level personally? How adept and qualified is this person to diagnose faulty performance and correct it? Thus we see that past performance and similar experience is encapsulated in every aspect of the program manager's credentials, those that accrue to the bidder's competency, or the lack of, in running a program.

The resumes of other managers, the technical staff, and support members are also of concern. It is known that some Navy proposal evaluators do not begin evaluating the resume from the top-down, but rather from the lowest salaried person shown in the resume section, working upward. If the project secretary answers the phone, completes and files the manager's routine reports, and coordinates for engineers and analysts, expectations of that employee's qualifications are pretty high.

If Configuration Management Technicians with a minimum of high school plus an A.S. Degree, three years of Navy equivalent experience, and a thorough, demonstrable knowledge of DOD-STD-100 Standard for Configuration Management, are specified by the RFP, then Logistics Technicians, fresh out of the fleet, had better not be proposed for those jobs. Medium-level technical staff, supervisors, and subtier managers will be subjected to the same investigation for related similar experience. The evaluators will fill out Contractor Clarification or Deficiency sheets on the resumes of each unresponsive resume, and, of course, a staff laden with enough faulty candidates may can be dismissed early for a comprehensive lack of experience. Navy evaluators have been known to screen the lowest level resumes first to get a sense of how well the proposed staff fits the RFP criteria.

Past Performance/Similar Experience in Deliverables. In most manufacturing contracts and in many services contracts, exact knowledge and experience in dealing with plans and programs, standards and specifications peculiar to the agency are essential in showing the client that relevant documents can be produced. Examples would include ILS Master Plans, Configuration Management Program Plans, Operations and Maintenance Plans, Overhaul Plans, Provisioning Spares Plans, Quality Assurance Plans, and the like, each tailored to the agency and to the program itself, at its unique geographical location, and with its own history and its own demographics.

Thus, a Quality Assurance Plan that was written for hardware products will be useless on a services contract for software maintenance. More important, the evaluators will rate this bidder as not just inexperienced, but incompetent, although the evaluation sheet or Clarification sheet may never say so. From a risk perspective, such a contractor gives the government the impression of a disaster waiting to happen—that is, a 100 percent probability that the contractor in question is grossly inexperienced in how to design an appropriate Quality Plan.

Thus, how contractors address their responsibilities to describe technically how they will respond to Data Item Deliverables (DIDs) and Contract Deliverable Requirements Lists (CDRLs) are, in effect, imminently traceable aspects of past performance and related similar experience.

The Federal Acquisition Streamlining Act (FASA) mandated the increased use of past performance as an evaluation factor in contract awards. In addition, the Office of Federal Procurement Policy issued detailed guidance for agencies in evaluating past performance. In many cases, government evaluators will seek to award contracts only to those contractors who have proven and unexcelled past experience in every aspect of technical performance. Whereas in the recent, past, Similar Experience and Past Performance was valued at as little as 5 to 10 percent of Evaluation Factors, FASA promises to routinely weigh in at 20 percent, or more, of every federal award. More important, the cumulative elements of past performance are spread throughout the evaluation criteria as filters to screen out imprecise, inaccurate, and misleading comparisons that the contractor may be portraying as exact matches to the RFP's requirements.

The following examples typify how evaluators may establish screens to assess for past performance in every proposal ingredient:

- Was the technical approach based on proven performance elsewhere, or just a hypothetical guess at what needs to be done?

- Were the contract references truly similar to the work specified? Were they near-misses, or completely dissimilar?

- Were the resumes responsive to the RFP's needs, skill levels, and client preferences? Were they incomplete or evasive?

© 1996 by Holbrook & Kellogg, Inc.

- Were costs consistent with substantial knowledge of the work? Were there significant costs included unilaterally to absorb the high price of a learning curve?

- Was the management plan based on techniques proven on this type of contract, and specific to these exact requirements?

- If live tests of equipment or demonstrations of systems were required, did the contractor show applied experience? Were the tests and demonstrations free from gross errors?

- Does the contractor's past performance equate to the RFP on a clear, concise, one-to-one level? If not, how close a match does the contractor provide?

- Are the deliverables based on completed plans, procedures, and other technical documents that were accepted before by a client?

Unless the answer to all of these questions is "yes," the contractor can expect to score poorly—that is, insufficient past performance and imprecise similar experience.

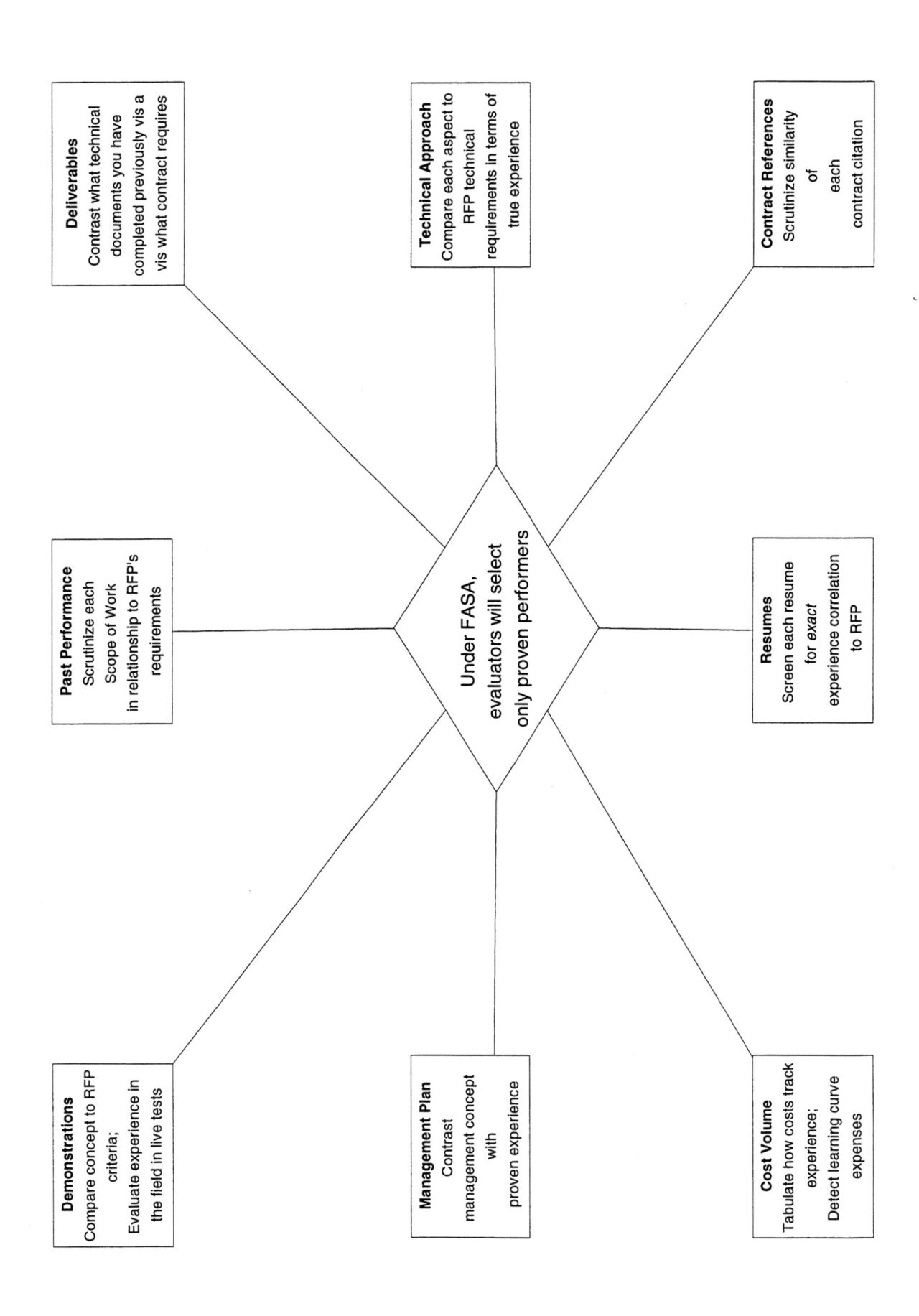

Deliverables
Contrast what technical documents you have completed previously vis a vis what contract requires

Technical Approach
Compare each aspect to RFP technical requirements in terms of true experience

Contract References
Scrutinize similarity of each contract citation

Past Performance
Scrutinize each Scope of Work in relationship to RFP's requirements

Resumes
Screen each resume for *exact* experience correlation to RFP

Demonstrations
Compare concept to RFP criteria; Evaluate experience in the field in live tests

Management Plan
Contrast management concept with proven experience

Cost Volume
Tabulate how costs track experience; Detect learning curve expenses

Under FASA, evaluators will select only proven performers

Figure 6.4
Get a Jump on the Evaluators—Scrutinize All Data for Exact Past Performance Matches

Implications of the Increased Emphasis on Past Performance

In mid-1995, sweeping changes were instituted in the federal government. One initiative placed greater emphasis in all agencies on past performance in awarding contracts. These changes were rooted in FASA and in guidance from the OFPP specifically OFPP Policy Letter No. 92-5, Past Performance Information. In brief, the goal is for all agencies to accrue accurate, up-to-date evaluations of every corporate entity that does or has done business with any U.S. agency. The likelihood is that the effort to develop governmentwide past performance information will continue throughout succeeding administrations, since the FASA database offers the government-at-large the promise of real advantages in evaluating proposals. If aggressively managed, a past performance database will offer every contracting officer and source selection official a yardstick by which to measure all contract performance.

In the past, some RFPs had assigned as little as 5 to 10 percent to Similar Experience/Past Performance, a formula geared to awarding contracts to bidders with excellent (if unproven) technical approaches. The OFPP direction subtracts points from other proposal evaluation factors and adds them to Past Performance criteria.

Today, an average federal RFP usually rates Past Performance as about 20 percent of the score, as a separate evaluation factor. The goal is to quantify how well a contractor has performed technically and managerially via documented achievements, milestones, and reports. More than likely, present and future RFPs will contain a separate Past Performance evaluation factor as well as another similar evaluation factor entitled Similar Experience. These are not double jeopardy, but they do have the potential to equal 40 percent of a proposal score, and can rapidly become the determining factor when evaluators cross-check the technical approach for Past Performance experience and weigh the Past Performance content of the resume section. In this case, Similar Experience is a reference check of how well the contractors' people have cooperated with their contracting officers, contracting officer's technical representatives and the agency staff at large, and as such is parallel as to the references people include on their resumes, references that are assured of giving good comments.

Firms that cannot or do not document excellence in past performance will rapidly experience punitive evaluations, especially in competitive RFPs where poor, checkered, or average past performance will be a good enough reason to omit a firm from the short list or make the award to a firm with better documented performance achievements.

Government Adjustments

The L and M Sections of RFPs (Instructions and Evaluation Criteria) will reflect the emphasis on past performance in future RFPs, elevating performance in effect over the technical approach, which was previously the deciding factor. If past contracts from an agency had a track record of performance problems, missed milestones, faulty products, and lackluster services, then past performance logically becomes the primary evaluation factor. Where the RFP deals with common support services, where the contractor's ability to respond rapidly and methodically is important, past performance is a good yardstick to gauge how well the contractor will perform. As the technical statement of work becomes complex and sophisticated, the contractor's past performance becomes the single indicator of what to expect in the next period of performance.

While OFPP recommends that past performance become the primary evaluation factor, the FAR does not support a universal shift. Thus, performance may be a subfactor under technical or management criteria. Past performance may crop up in RFPs in a number of ways. Past performance may be combined with corporate experience. Performance can then be readily linked to and evaluated with quality overall—that is, not just that a firm has done a type of work before but how well they did or did not perform their overall responsibilities. Another technique is the addition of performance as a risk factor within the evaluation factors for the management and technical approaches. In this case verified performance records will be studies both objectively and subjectively for elements of risk. Where the contractor's goods or services failed tests, caused accidents, or had an unsatisfactory trend analysis for recurring problems, the government will expect future troubles based on past performance. Matters of record, such as a contractor's inability to comply with standards and specifications for materials or workmanship are evidence of imprecise understanding, and may lead an objective evaluator to expect future shortfalls. If the agency has placed numerous Memoranda to the File in a contractor's contract file, it is evident that client satisfaction is lacking, thus establishing a history of inattentiveness to the goals of the end user, i.e., the agency that is paying the bills. The annual, quarterly, and monthly reports submitted to the government (and usually obtainable through the Freedom of Information Act) document the contractor's ability to

- plan and schedule;

- get productivity out of their employees;

- forecast and contain reasonable costs; and

- complete or make progress on technical directives from the customer.

All of these documents accrue to how past performance will be seen when evaluated in competition.

The new emphasis on Past Performance will work a hardship on new firms who have little or no contractual experience to draw upon. Typically, these firms will be small businesses sponsored by the Small Business Administration and guided through noncompetitive or limited competitions where full-scale RFP evaluations are not conducted. The government is supposed to regard their experience as a neutral factor, hardly an incentive to compete against larger firms with substantial, positive success stories on related work. However, new contractors can gain past performance credentials by serving as subcontractors in their own areas of expertise, in time growing into related areas. Another way to gain acceptability and recognition is to include the resumes of the small firm's principals, where those individuals have played a technical or management role, at a larger firm in achieving quality in past performance.

How to Address Past Performance Problems Positively. Since there is no perfect contract, there can be no perfect contractor. But some approach perfection. They react positively to criticism, practicing proactive management, intercepting most problems in the formative stages. Here are some ways to address problems positively in Past Performance sections:

- Try to limit the number and severity of problems on the contract. Move rapidly to contain and remedy bad situations. Don't let them balloon into a series of negatives. Where an event occurred that was beyond the Scope of the Work or beyond the contractor's control, records must reflect the reasons for the event and what exactly transpired. Where the performance is excusable, the government must be convinced that there was no serious error, omission, or neglect. Eye witnesses and accurate recordkeeping will protect the contractor from unfounded allegations.

- When writing a description of a contract that had a problem, write from the positive angle of "Lessons Learned," documenting how the contractor and the government

    ▸ apprehended and grappled with the problem;

    ▸ broke it down to its root causes;

    ▸ acted aggressively to treat every symptom, and

    ▸ took corrective action.

    Often, the government's or contractor's Quality Assurance office will write a QA Deficiency that requires a Corrective Action Report (CAR) be written to describe how a problem was fixed. These become a permanent part of the contract file, accessible to competitors. Unresolved or partially unresolved CARs give the Inspector General, the DCAA, and the GAO the impression that the contractor will not or cannot correct a deficiency, and hence is a candidate for replacement.

- Be sure you know where the evaluators got their performance data. Although the government and commercial sources of information aim to be error-free, those databases are in the formative stages and in flux, therefore subject to flaws. If you suspect unfair criticism has entered the evaluation process, putting your firm at a disadvantage, it is appropriate to ask the firm's legal and contracts staff to put together an inquiry or protest, asking the evaluators to divulge the sources, reliability and verifiability of their data.

- Write Past Performance sections to proposals from a risk-reduction perspective. Focus all the company's technical and management activities on 100 percent compliance to the Scope of Work, end user/client satisfaction, and Zero Tolerance for sloppy workmanship, nonproductivity.

    ▸ stress timeliness, milestones, and deadlines;
    ▸ equate costs to meeting performance goals;
    ▸ stress quality consciousness throughout.

© 1996 by Holbrook & Kellogg, Inc.

Without accurate, timely records, the contractor's own assessment of the work will be viewed as a public relations story. Keep copious and detailed records to substantiate good performance.

Proactive management of every contract is the best curative for writing Past Performance and Similar Experience narratives. Aggressively solving the client's problems throughout the contract is much more observable, consistent, and laudable than an eleventh-hour or Fire Department style of attempting to cure a poor performance record late. Firms have phrases for these late-fix scenarios. These include: "Putting On A Happy Face," "The Turnaround," or, last and least flattering, "Putting Lipstick on the Gorilla." None of these are particularly advisable because there is no continuum of success stories from Day One of the contract until today, no kudos for the staff, and no particular recognition of the program manager and senior staff. In other words, here's a thoroughly average, nondescript contractor, ready to be replaced by someone whose credentials sparkle.

Add to this the new FASA implication that a central federal database will house *all* references on *all* contractors, and that commercial solvency and character databases at Dun & Bradstreet (D&B) will be used to validate contractor claims, and we have an exciting new era to explore. D&B's database in particular may de-bunk contractor assertions that they offer an unblemished record of performance. The client can place a 5-minute, $50.00 telephonic investigation request to D&B that will yield

- the full legal names, addresses, and personal histories of all the principals,

- any local, state, federal, or civil judgments made against them by any court,

- delinquencies in paying subcontractors and consultants, and

- liens from landlords and computer or furniture stores.

Prudent prime contractors will want to assess the current reputation of small or disadvantaged firms before adding them to a proposal team for the same reasons the government's evaluators are doing so.

CONCLUSION

Contractors' need to address past performance and relevant similar experience is required throughout the proposal. They need to demonstrate in as much detail as possible, and do so verified by the client. The proactive management approach begins as soon as the contract is awarded, in the so-called Post-Award Euphoria Days when the contracting staff is relishing its own professional competence in choosing that bright, new contractor and turning out the uncooperative bunch of rogues who used to wrestle with performance. As the government managers work with their contractor equivalents, the contractor assures good will, attention to details, and the actual planting of kudos where the opportunity exists. This continuum of good will typically will cushion and encapsulate the work-a-day world of the three to five years of the contract, insulating contractor and government management alike from project problems, cost overruns, project lateness, and the like in which the contractor supports and counsels the client in a businesslike and proactive manner, not allowing themselves or the government to be subjected to unjust criticism without rebuttal and the chance to respond with a Plan of Action and Milestones. In this manner, the Similar Experience or Past Performance section is

- writing itself,
- accumulating letters of praise,
- documenting and resolving contractual problems,
- hiring (and firing) prudently, and
- growing the contract in directions suggested by the client.

6.5 WRITING THE RESUME SECTION

As we have discussed in Similar Experience/Past Performance, every customer is extremely concerned about the quality of personnel who will be assigned to a program and about the contractor's ability to perform the work well.

In Similar Experience/Past Performance, we stressed that those sections must be written with care in order to maximize the company's skills and capabilities in the eyes of a particular client. In the resume section, we again focus on emphasizing skills and credentials, but in this case, for the skills and credentials of individuals as they are shown in corporate resumes.

© 1996 by Holbrook & Kellogg, Inc.

It has been said that in the services industries, the people are everything. That is, the client has to interact and cooperate with the contractor's staff so that staff has to be compatible with the client ideal of its personnel needs, knowledgeable of the contract's parts, recognized as appropriately educated, properly experienced, and as intimately familiar with the project as humanly possible. This is a tough bill to fill. Contractors often overlook the Resume Section until late in the proposal when it may be difficult to make changes, and substitutions for resumes are deemed unsatisfactory.

6.5.1 UNDERSTANDING HOW CLIENTS SCORE AND EVALUATE RESUMES

In federal RFPs, the L Section describes Instructions for the proposal, usually specifying a resume section and asking for resumes of Key Personnel and others that the contractor may wish to include. The M Section of the government-style RFP then spells out how the evaluators will score the Resume Section. Contractors are well advised to expect that evaluators will scrutinize resumes as thoroughly as any technical or management aspect of the proposal.

A common checklist evaluation of the resumes will include taking the staff in what could be described as reverse order of importance—that is, checking the credentials of the hourly and junior salaried personnel first. Scrutiny will be devoted to assure that each person meets without interpretation the RFP's criteria for employment. A typical Navy RFP might require that for a particular contract an entry level Configuration Management (CM) Technician will have a high school diploma, or equivalent, one year in a particular tech school and with on-the-job training, selected CM job tasks, and three or more years in the field with Navy or industrial CM experience. That's the minimum, the floor, the baseline—nothing below those qualifications counts. Evaluators who find an ILS Technician here with two years shipboard CM experience have no choice but to write a Deficiency or Clarification that counts against the contractor's resume score. Too many Deficiencies or Clarifications means the contractor is nonresponsive in personnel or does not know how to select simply and hire the right project personnel.

Moving upwards through the Resume Section, evaluators will again closely match credentials to the RAP's requirements, writing Clarifications and Deficiencies everywhere the information does not stand alone plainly. In mid-level resumes, where supervisors and engineers usually appear, evaluators will be looking for project-specific knowledge. In fact, they will be seeking parallel or identical experience and exact correlations to the Statement of Work. Without solid cross-references, evaluators tend to hold resumes as generic, second-rate statements of credentials.

The management tier, especially the program manager's resumes, get special attention and the scrutiny is intense, delving into a person's career for 5 or 10 years and further. With the management people, the client expects to find something really special, a nearly exact match of people to problems and of contractor management personalities to client management personalities. Consequently, the management tier should have been introduced to the client prior to release of the RFP, perhaps with an informal meeting or get-together such as a contractor's open house, a professional meeting, etc. This will preclude a retired Army officer being proposed for a contract where 10 years' Naval shipboard experience is strongly recommended. Reputation and compatibility are also at issue: Can these people get along with this client for five years? Does the top contractor manager complement and enhance the client's management framework, or is that manager a perpetual sore point?

The proposal manager should read and study all resumes at three stages, in the beginning when a Key Personnel list is outlined, at mid-point when the resume team has had an opportunity to assess the availability and qualifications of the staff in mind, and after Red Team, when responding to critiques about the credibility of the resumes. Whenever the proposal manager or the resume section leader suspects a resume is faulty or deficient, that resume should be immediately improved or replaced by a better one, until the entire resume section is assured of a near perfect client score.

6.5.2 AVOIDING BAIT-AND-SWITCH ACCUSATIONS

As resume teams collect, select, and polish resumes, they should remember that reference checks should be made within the company to assess if the firm's best talent is being shown and that there is a reasonable probability that these people will in fact be available when the award is made.

It is incumbent on the contractor to put forward the best team possible, remembering that the client expects to see and have working on its project many, if not all, of the persons whose resumes appear. Availability of people depends on the client's award date. Amendments usually delay award and increase the possibility of losing personnel to other projects. The client has a right to know that information, but only if they ask; do not unnecessarily alarm the client with imprecise news. Most agencies allow for some substitutions to be prudently presented as upgrades, but these must not appear to be "Bait-and-Switch" accusations, where the client implies that downgraded, problematic employees will appear after award, and not the superb team shown in resumes.

The client identifies persons that they feel are unqualified via Clarification and Deficiency questions, and probes the credentials and availability of certain persons. Substituting overqualified people to replace such persons can lead to a "Bait-and-Switch" accusation with the client suggesting that the firm really has no intention of providing such talent at the salary levels the client can afford.

When the client specifies percents of time to be devoted to a project, it is advisable for the firm to indicate 100 percent commitment of staff, possibly listing the program manager at 95 percent to explain some corporate involvement. When firms nominate an officer of the corporation to serve as a program manager, it is not credible, for example, for the president to promise 100 percent of his/her time to the contract, knowing full well that the CEO cannot spend more than 50 percent of his/her time on any project. Credibility is an issue with most clients, and the client has a tendency to make worst-case assumptions, putting the contractor on the spot to prove every promise.

Availability of personnel has become a sore point with the government since 1992, and many RAPS now require signed, dated resumes upon proposal submittal. More commitments such as letters, with salaries agreed to, must be included in the cost proposal. Avoidance of these commitments will be regarded as bad faith, and cause for not being selected.

The resumes must be signed and dated and accompanied by commitment letters. The proposal manager needs to assert control on Day One of the proposal, have a strong and aggressive resume team leader on hand to select, and groom the credentials of people who are qualified and, in fact, available.

6.5.3 WRITING THE TECHNICAL/BUSINESS RESUME

Like the rest of the proposal, the resume section needs to be written to match the RFP. In point of fact, however, many firms whose business is technology have neither an orderly technical/business resume file or know how to use one. Some rely on the resumes that employees were hired from, which is asking for the employee's credentials to be on record in an outdated and inaccurate form.

Most people equate a resume to their own personal employment, so for definition purposes in this section we will differentiate *personal employment resumes* from *technical/business resumes*. The personal resume was written to get an interview and land a job for *that person alone*. The technical/business resumes are by definition written to get employment for *a team of people* through a company on a contract basis. Personal resumes tend to be broad and generic to generate multiple interests, whereas technical/business resumes tend to be much more exacting, precise, and project-related. The technical/business resume is aimed at showing teamwork and group strength, where and how the engineer or technician will fit into a proposed effort.

Most firms take custody of the resumes of new employees and consolidate and format them in a company-specific manner. The goals of this transformation should be to tailor the resumes to the clients' expectations, deleting excessive personal data and focusing on skills and abilities that will score well. The formatting of so many resumes raises the key question, "Is there any one best resume style for proposals, and if so what are the characteristics?" We show the five predominant types of resumes encountered in American industry today.

- The Telegraphic Resume
- The Scholarly Resume
- The Categorizing Resume
- The Job Application Resume
- The Balanced Resume

They are shown in Figures 6.4 through 6.8, following.

After comparing and contrasting various aspects of each of the five types, it will be readily apparent that the most workable technical/business resume is type five, the Balanced Resume, a type of document that is prepared with special emphasis on each RAP and the technical requirements of that client. In discussing the aspects of this and the other resume types, we use plus signs (+) and minus signs (-) to show the impacts of what is discussed upon the reader/evaluator. We also provide the writing remedies to arrive at more persuasive technical/business resumes.

© 1996 by Holbrook & Kellogg, Inc.

Certain client RFPs discourage or do not allow the contractor to provide independently derived resumes. Instead they provide a specific format and contents, and a 1-2 page limit. It is imperative that contractors regard such RFP requirements as binding and recognize that deviation from resume specifications will cause technical disqualification. Such RFPs require the ultimate in resume tailoring, selecting, emphasizing, and condensing areas of technical expertise. Each of the following resume types provide some guidelines for the resume writing team to keep in mind, but the most important inquiries to/about another person's resume come from close analysis of each resume, then visiting or calling each person when a question arises. There is no substitute for personal knowledge of another person's resume. It is often advisable for the resume team leader to plan and schedule, at the outset of the proposal, short meetings with selected people to update, focus, and tailor their resumes.

The Telegraphic Resume

According to one school of thought in resume writing, "Less is better." Resume writers that follow this advice condense a person's entire career into a few brief statements (see illustrations), with an entry for education and some attention to similar projects. A tight format is used to limit the data. These resumes are usually too brief or truncated (-). The person comes across to the reader as a stranger. These telegraphic resumes are restricted to the blocks in the format (-) and often have a lot of wasted space (-). However, telegraphic resumes do have one big advantage: They were prepared just for this proposal (+), eliminating a lot of obsolete, inappropriate data that the client probably would have glossed over or regarded as eye-wash.

In 1994 and 1995, some RFPs from commercial firms seeking to outsource services were specific and demanding in their instructions for very limited resumes, much like the Telegraphic Resume, so brevity and specificity may be a wave of the future. They are, of course, easier and faster for the evaluators to read and consequently easier to score.

The Scholarly Resume

Almost everyone who has written or reviewed proposals has had to endure some Scholarly Resumes, those overwritten, and extremely long documents that tend to be used to show scientific expertise. The scholarly resume consist of an extremely lengthy account of each activity this person has been engaged in throughout academia and his/her career, followed by a copious descriptive bibliography of all the reports, deliverables, papers, and professional publications that he/she wrote, edited, or participated in. Six or seven pages of data are not unusual for the scholarly resume where the rule is "More is better."

Some firms that perform a lot of original research and development, highly scientific assessments, and employ many earned PhDs want every resume to be this rich in detail and description, without concern for the evaluators who have to read them (-). Firms whose line employees are not research scientists have adapted their resumes to a scholarly format in an attempt to simulate a great deal of in-depth knowledge of a technical field or fields. All too often, such resumes restate old Scopes of Work, project specifications, or contract requirements (-), making them dense reading. Reading is further complicated with the first person narration, "I did...," an intensely personal view (-).

Generally, these scholarly documents (or Academic Resumes as they are often called) are partly long paragraphs and partly lists of things. The paragraphs are encyclopedic and dense in scientific meaning; the lists are bibliographies in disguise relegated into what looks like narrative reading on the first encounter. These resumes often are presented in reverse chronology (-), asking the reader to begin with a person's oldest job first, then work up to more recent assignments. Scholarly resumes often appear on plain bond paper, with journal articles or abstracts stapled to them.

These resumes are too long (-) for most clients to pay much attention to and too tedious to make a good technical impression (-), and then there is the matter of the reader's suspicions being evoked: Isn't there a lot of boilerplate padded into these resumes with the clear intent to embellish the ordinary? Also, many such documents appear to ask the evaluator to be a scholar of the particular discipline, a dangerous assumption, especially when contracts reviewers at federal agencies may be the intended audience. At least the Scholarly Resumes have one benefit, depth (+), which is an asset for the resume writing team to draw upon selectively. Government-style SF-171s are in the same camp, with a vast array of data instead of a selective presentation.

When interviewing people about their resumes, the people who drew up their own Scholarly Resume or Academic Resume will be far and away the most difficult to discuss their careers. (With a substantial number of Ph.D.-type personalities, the idea of reducing their own holy writ to some demeaning technical or business purpose is

© 1996 by Holbrook & Kellogg, Inc.

heretical). It is essential at the onset of the proposal that the proposal manager inform all staff members that the company owns the technical/business resumes, that it is not their personal property, and that the resumes will be formatted accurately and expeditiously to place the company in the best competitive light. On the scientific and R&D side, of course, there are ample examples of RAPS where the client actually asked for these lengthy resumes, which may mean that an occasional engineer or analyst may have to fill in details to parallel a bulky format.

The resume team leader will usually have to remind people that only the proposal manager has final authority for what goes into the proposal resumes, period. By handing out a prescriptive Sample Resume early, along with paragraph- and page-length guidelines, the proposal manager and the resume team leader can save themselves a lot of extra work when handling scholarly, academic resumes. If the resume team has to take responsibility for editing such resumes for content and format, then they can do so with uniformity and consistency.

The Categorizing Resume

When technological firms reach maturity, they obtain a firm grasp of their marketplace, client expectations, and how resumes develop the client's estimations of the firm. At that crossroads firms often make a concentrated effort to clarify and sophisticate all of their resumes. The degree of self-esteem that the firm feels usually dictates the budget that they will dedicate to get a companywide collection of standardized categorizing resumes. A mandate and example (sample) will be generated and issued to each employee, with a deadline. Supervisors will screen drafts to get consistency. Sometimes called "The Resume Drill," this update of the entire firm's resumes is aimed at embellishing and enhancing the entire career of each employee. Photograph sessions are often scheduled to get flattering portraits of each team member, and even the clothing (suit, tie) are mandated for uniform businesslike appearances. The photos are then set into the resume format, which may be printed in two colors, with prominent company logo.

The categorizing resume has an ornate format that is easy to read but hard to write (-), since there is the urge to force everyone's experience into exactly the same slots, with a few sentences explaining the categories of information. The third person pronoun is carefully adhered to. The format can be too loose and waste space (-) or have blank spaces; other categorizing resumes cram all data onto one sheet, creating the impression that the firm doesn't care about readability, margins, consistency, or the loss of emphasis (-).

The great asset these resumes offer is dedicated efforts to update and amend them (+). However, they are often outdated 90 days after they are finished. These categorizing resumes should always have some type of code on the last page that decodes to the date of preparation, or last revision effort. Typically, firms that use categorizing resumes also ask all employees to update/upgrade their resumes prior to their annual performance review thus automatically assuring that career employees will have strong, current resumes.

In summary, the categorizing resume has some things in its favor, with few disadvantages. Primarily, the categorizing resume has a fully formal appearance (+) and always looks current (+), whether it is or not. The categorizing resumes are expensive to produce and consume much more time than firms estimate (-). Their great strength is that it has the aura of extreme organization (+) and thoroughgoing professional discipline (+). If the senior management of most firms had an entire day to reflect on the impact that resumes make on evaluators, it is likely that businesses would almost universally demand that their proposals contain only categorizing resumes, irrespective of cost and efforts.

The Job Application Resume

At certain stages in a company's development the firm grows too quickly or loses sight of its marketplace and the uniqueness of its technical people. Then, the business/technical resumes suffer. At first, some manager or the company's publications group may try to update and standardize resumes, but without regular attention the efforts fail and groups of technical employees branch out into a rainbow of resume formats. At that point it's everybody for themselves. Each engineer, architect, or analyst will tend to revert to that personal, job application resume that won the interview and got the job offer. At this stage, even the best employees will confuse the personal resume with a corporate technical/business resume.

The hallmark of the personal or job application resume is the clutter of denominational, political, or hobby references with the technical world (-). These resumes overplay or underplay assignments, roles, or skill levels—their own or the company's. The narrative mixes "I did's" with "They did's." A single format will not emerge. Fonts change, categories of information vary, and old, odd formats will show up, perhaps with some typeset text and taped-in photographs.

Imagine the plight (or amusement) of the evaluators who reads a proposal containing a bevy of such resumes, all typed on different machines, with bizarre margins, and a potpourri of formats! The materials are guaranteed to be from prior jobs (-) once done and forever unchanged. These documents appear on plain white paper and are almost worse than no resume at all. They are neither specific (-) nor up-to-date (-) and show that management has little regard for control of its people (-). Thus, the job application resumes are like a team of job candidates pretending to be a corporate team, except that when they speak, they interrupt and contradict each other.

Some firms have a bad habit of adding nonspecific (job application) resumes at the end of a resume section, as though these odd, outdated variety of vita sheets would add some value in the mind of a skeptical evaluator with a precise resume checklist to run. An otherwise cogent, standardized set of resumes could thus lose points via the careless addition. This afterthought of "resume dumping" should be avoided by proposal managers who remember that the evaluators have the last say (and the last laugh) at resume sections.

The Balanced Resume

After examining the other categories of resumes, only one type emerges as best-suited for most RFP uses. Because the Balanced Resume incorporates the best traits of the others while excluding their problems, this type can be recommended to the resume writing teams of virtually any firm, large or small. From the Telegraphic Resume, the Balanced Resume brings a specific tailoring just to a specific RFP. Data exactly pertinent to the RFP will be emphasized early (+) and catch the evaluator's eye. Brief explanatory sentences will precede lengthy lists so that raw data becomes rapidly significant. Insignificant materials will be edited out. Paragraphs will be seven or eight sentences at most. No resume will exceed two pages in length, demonstrating reader friendliness (+) and the ability to communicate effectively and concisely.

From the Scholarly Resume, the resume writing team will take selected technical accomplishments and publications, focusing on areas targeted from the RFP (+) that show extraordinary expertise. There will be no attempts to "snow" the evaluators or overwhelm the reader with uncorrelated facts or a superfluous flood of information. Lengthy Latin phrases will follow a low ballistic trajectory into the wastebasket.

The Categorizing Resume brings a businesslike appearance to the Balanced Resume, with its bordered bond paper, logo, photo, and latest, most persuasive data for each entry (+). For its own organization, the Balanced Resume has only three categories—name, experience, and education, broad enough to contain the most diverse career data. Since evaluators care most about experience, the Balanced Resume gives experience first, giving a good interview (+) even though the evaluator may be hundreds of miles away. This resume starts with the most recent, relevant data, reserving some room for older work. In this education block, all professional seminars, degrees, courses, certificates, and work in progress are appropriate. Third Person is used throughout, with none of the idiosyncrasies of the Job Application resume.

If stored on the hard drive of the firm's server and held separately by the resume team leader, the Balanced Resumes are constantly available to users to update and keep current. Depending on graphics capabilities, photos may be routinely dropped onto resumes without the efforts needed in the early 1990s.

In summary, the Balanced Resume alone speaks like a confident person at an interview, articulating smoothly why this candidate is best-suited to the job, building on similar or exact skills and expertise, and matched to the client's priorities.

PROFESSIONAL DATA	Employer's Company/Name
	Resume Data RFP 123456789-96

John Doe
_____
Name

Senior Electronic Design Engineer
_____
Title

E-10
_____
Pay Grade/Level

90% - 100%
_____
Percent of Time Devoted to Program

CURRENT RESPONSIBILITIES

Nature of Program

Now PM of S-Band Radio used by USAF on ALCM, 1993-Present. Brief summary of technical aspects of of this project.
_____

_____

_____

SIMILAR EXPERIENCE

Former Program One

PM XYZ, S-Band Radio, 1991-1993. Description of project specifics, briefly discussed.
_____

_____

Former Program Two

Asst. PM, ABC HF/VHF Monitors, 1988-1991.
_____

_____

Figure 6.5 *Telegraphic Resume*

© 1996 by Holbrook & Kellogg, Inc.

Former Programs—Other

Technician/Drafter, GHI Radio Sets, 1983-1987.

EDUCATION

Degree

UNIX, El Paso, Texas, B.S. in Electronic Engineering

Advanced Degree

15 hours--Master's level electronic design

Courses and Seminars

ANSI Seminar on Circuitry, 1989

**LICENSES, REGISTRATIONS, PATENTS,
BUSINESS AND PROFESSIONAL CREDENTIALS**

Member, IEEE

Member, U.S. Army Reserve--Electronics Company

Army Electronics Schools

Figure 6.5 *Telegraphic Resume*

© 1996 by Holbrook & Kellogg, Inc.

JANE DOE, SENIOR ENVIRONMENTAL SCIENTIST
Hypothetical Resources & Environmental Study Corporation

Project Director for Biological and Botanical Species and Habitat Analyses for North America and Portions of South Central America

EDUCATION: B.S. - 1988   M.S. - 1991   Ph.D. - 1995

B.S. Degree, Botany, Massachusetts Institute of Technology

Areas of Emphasis: Deciduous Trees of Northcentral U.S., et al.
Minor Emphasis: Biology, Worm Phenomena in Northcentral Forests, et al.

M.S. Degree, Botany, University of California at Los Angeles

Principal Studies: Electron Microscope Analyses of the Microstructure of the Deciduous Trees of the California Desert Regions; Interaction among Annelids and Nemerteans in Tropic and Subtropic Biospheres; Biological Aspects of the Desert.

Thesis: A Study of Larval Metamorphosis of the Scutigera Coleopatrata in an Atmosphere of Juniperus Virginia.

Ph.D. Degree, Environmental Life Sciences, University of California at Berkeley

Principal field of study: Interaction of Biospheres of Western States in both closed and open botanical and biological habitats.

Dissertation: An In-Depth Computer Simulation of the Environmental Impacts of Major Bridge Construction Projects on the Ecosphere of Northern California, Its Streams, Rivers, Watertables, and the Flora and Fauna Thereof.

EXPERIENCE: California Air and Water Authority, Orzona County, 1995 - Present

Responsible for investigations into flora and fauna of Orzona region ecosphere of the Argiope Aurantia species and impact on Citrus Sinensis. Projected species duration and endurance under the effects of dichloridphenyltrichloroethane, under tropic and subtropic atmospheric conditions. Position was Science Intern. Took Post-Graduate Courses in Advanced Perca Flavenscens Species. After completing this internship, under direction of the Senior Air and Water Scientist, became project scientist for the county's river and stream studies where responsibilities centered on the biochemical ramifications of hydrocarbon vapors and gaseous and liqueous masses when infused in flumenary bodies.

EXPERIENCE: Pre-Hypothetical Resources & Environmental Corporation, 1988-94

After joining PHR&EC in 1988, Ms. Doe was a Division Group Scientist Level I, responsible for

(PHR & EC Continued on Resume of J. Doe, page 2)   (J. Doe Resume - Page 1 of 7)

Figure 6.6 *Scholarly Resume*

© 1996 by Holbrook & Kellogg, Inc.

```
┌─────────────────────────────────────────────────────────────────┐
│  O Company Logo, Formal Layout Design      THE MAJOR ARCHITECTURE CO., INC
│
│                        ┌─────────┐
│                        │         │
│                        │Portrait of│   (All Text Typeset)
│                        │ John Doe │
│                        │         │
│                        └─────────┘
│
│                        MR. JOHN DOE
│                   REGISTERED PROFESSIONAL ARCHITECT
```

COLLEGE	**YEAR**	**DEGREE**
Illinois State University	1979	B.S. in Architecture

PROFESSIONAL CREDENTIALS Registered Professional Architect—Illinois, Indiana, Ohio and Pennsylvania

ARTICLE, JOURNAL OF ARCHITECTURE
"Metropolitan Stadia Planning," July 4, 1989, Pages 67-71

Management for Architects and Engineers, Seminar, 1991

PROFESSIONAL EXPERIENCE

1979 - Present   *THE MAJOR ARCHITECTURE CORPORATION, INC.*

Mr. Doe began his career at MAC as a Project Engineer in charge of Structural works for CSS Rework Project, 1979 - 1984. He was responsible for a team of architects, designers, drafters and technicians. Based on success of that project, he was promoted to his current position in 1985, overseeing all MAC stadia programs.

Other major reworks and evaluations of stadia were conducted by Mr. Doe for MAC clients in municipal governments in Texas, Maryland, Oklahoma and Ohio, 1985 - Present.

(Page 1 of 2 - Resume of John Doe, RPA)

Figure 6.7 *Categorizing Resume*

© 1996 by Holbrook & Kellogg, Inc.

RESUME FOR JOHN ANTONIO DOE, P.E.
1437 Avenue Northwest
Some Town, New Jersey 00000
Telephone
(000) 000-000*
*After 5PM, please

Doe, J.A.
Pay Gr. 5A
Civ Engr
Iowa Roads
Proposal

Born:	October 31, 1958 - Chicago, Illinois
Married:	Wife: Jane
	Children: John Jr. and Joann, ages 17 and 8
Primary School:	Decatur, Illinois Public Schools
Junior High School:	Chicago Public School #47
High School:	Chicago Public School #61
Clubs/Organizations:	Knights of Columbus, IEEE, ANSI, Elks, Masons, Friars, Area United Way, Senior Campaign Official VFW, Church Attender

COLLEGE EDUCATION

Graduate Engineer from Ohio State University in 1982 with Bachelor's Degree (B.S.) In Civil Engineering. Earned 70 percent of tuition. EIT and RPE Licenses in Ohio, grandfathered in Pennsylvania and Illinois.

Minor in American History, emphasis in Civil War and Economic Impacts on Modern U.S. Society. Member, College History Debate Club. Member, Young Progressive Party.

WORK EXPERIENCE

Industrious worker, employed since age 14. First job as night watchman for uncle's watermelon patch. Since then, moved up the ladder progressively by working at gas stations, grocery stores, to summer jobs in civil engineering construction crews, pipeline and sewer line construction projects, 1978-1981. Worked in civil firms while in college. Conscientious, dedicated to my work.

MID-STATES HYPOTHETICAL CIVIL ENGINEERING FIRM, LTD., 1982 - Present

Hired by J. Smith to perform civil engineer work on roads to timber sites in remote areas. Planned 300 miles of routes of gravel roads, graded them, set up gasoline stops. Field Engineer. Built the roadways. Later work on channels and causeways in Illinois. Promoted to Chief Engineer when J. Smith left the company in 1984. Wrote reports, specifications, purchase orders and letters. Passed EIT exam, 1987. Passed Registered Professional Engineer's exam, 1989. Expert witness, C.E., Ill. vs. MSHCEF Ltd., 1994. Extensive project work.

SMALLER LARGE CIVIL ENGINEERING, INC, 1978-1982

Hired by J. Jones at substantial salary increase over summer work. Better career opportunity. In charge of road jobs, new civil engineer technicians, surveyors, and construction crew. Developed SLCEI's first CPM planning strategy, joined Toast Master's Club, attended professional seminars for Mac and PC/DOS CPM approaches.

Figure 6.8 *Job Application Resume*

© 1996 by Holbrook & Kellogg, Inc.

JOHN DOE
Senior Petroleum Engineer
Large Oil and Gas Firm (LOGF)
Tulsa, OK

CAREER HIGHLIGHTS

- *Use this block to highlight and identify special skills and experience related exactly to this client's scope of work.*
- *Refer to deliverables, plans, and hard accomplishments that were this individual's responsibility.*
- *Cite performance-related work efforts, chronologically.*

EXPERIENCE

Mr. Doe is the proposed Manager of Petroleum Engineering for the XYZ Program for the ABC Production Project. He has over 10 years of expertise in oil and gas fields in the areas of production, well sinking, lining, and connecting pipelines.

Mr. Doe's current assignment is in the Gulf of Arabia on a major field development job at a site which promises to yield 6 million barrels per day within the next few months, plus two million barrels per day of gas flow. He supervises a team of six petroleum engineers, two geologists, seven draftsmen, seven technicians, and a field staff of some 50 locals and company personnel engaged in actual drilling operations.

As a Senior Petroleum Engineer at LOGF, Mr. Doe is responsible for overall operations at the site. He administers, negotiates, and manages all properties, equipment, and people at this location, at $4 million per year contract obligation. After his present assignment winds down in three to five months, he will be made available to head the XYZ Program, which is almost entirely similar to his Gulf of Arabia program.

Prior similar work includes Senior Petroleum Engineer and District Manager of the following sites: Indonesian sweet crude project, 1983-87; Gas Development Project, 1988; Abu Dhabi and Saudi Arabian Petroleum Development Venture, 1989-present.

Prior to coming to LOGF, Mr. Doe was responsible for general engineering and property management, reservoir and reserve studies for the Minor US Oil and Gas Co., Ltd. He is an expert in foreign and domestic properties and leases. Mr. Doe has also done extensive logging, design, initiation, installation, and operation of secondary recovery projects.

EDUCATION

B.S. Degree in Petroleum Engineering, 1983, University of Tulsa.

M.S. Degree in Petroleum Engineering, 1986, University of Tulsa.

Advanced Petroleum Reservoir Engineering Studies from the University of Oklahoma, 1990. Registered Professional Engineer in Oklahoma, Texas, California, and Louisiana. Member: AIME and AAPG. Four articles in *Oil and Gas Journal*.

Figure 6.9 *Balanced Resume*

© 1996 by Holbrook & Kellogg, Inc.

Summary of Technical and Business Resumes for Proposals

As we have shown in the sample resumes, in industry today you find a wide variety of resumes. The proposal manager should assign a resume writing team to gather, edit, and tailor the resumes to the requirements of the RFP, irrespective of the personal opinions that contributors of resumes may have about their own resumes.

The technical/business resume is the property of the company and can be shaped and emphasized to win particular contracts, and the proposal manager is the final arbiter of what and who is ultimately presented in the proposal's resume section. However, the organization chart that appears in the proposal must contain all personnel designated "Key" by the RFP. Therefore, preparing a resume for each "Key" person has to be done with precision, designating how each such professional or technical employee makes exact contributions to the contract. Important nonkey employees who will be assigned to the contract also deserve a through, professional resume, however, these resumes may have less time and energy invested in them than the "Key" and manager level resumes.

The recommended approach is for the proposal manager and resume team leader to select promising candidates by resume credentials, investigate and polish each resume to its ultimate end product, and lastly to decide whether to include each resume or opt for someone else. Every selection will be based on the RFP's evaluation criteria and instructions, (RFP Sections L and M), plus whatever marketing intelligence can be useful about the contract in question. More technical RFPs often contain the client's Job Descriptions for the personnel they seek—the proposal manager and resume team leader have to be 100 percent certain every resume that they submit meets or exceeds the requirements, or the client's evaluators will write a score-lowering Deficiency or Clarification inquiry directed at personnel staffing. Candidates should therefore be called or interviewed in person by the resume writing team when there is any question about a resume's dates, longevity by experience, background, qualifications or credentials.

The Statement of Work (SOW) of an RFP contains key words and criteria for selecting personnel for the project. The resumes chosen should link closely to the client's key, central concept of how the work will be performed. An impartial reader, such as the proposal manager or Red Team member, should be able to pick up any resume—or all of them as a team—and readily see the person's role, usefulness, contributions and interactions with the other "Key" staff members. Unless the program manager's own resume reflects all the aspects of the SOW, it is not complete. The program manager's resume must show total coverage of the SOW, and expertise into all key facets, with supervisory oversight of supporting roles.

© 1996 by Holbrook & Kellogg, Inc.

CHAPTER 7

The Proposal Manager

It has been said that the proposal manager is the combination of a ringmaster at a three-ring circus and a slave driver. There is some truth in both views. The proposal manager certainly stamps his or her personality on the proposal. In brief, the manager is responsible for:

- Impressing his/her own staff as a first-rate project manager;

- Overseeing proposal progress from start to finish;

- Selecting the right deputy proposal manager;

- Calling together and naming volume managers, and for selecting separate, specialized technology writers;

- Making sure all interactions with the proposal development organization flow smoothly;

- Managing the writing plan for product quality, schedule, and cost;

- Acquiring or disposing of personnel who can help generally, specifically, or not at all; and

- Personally editing the document so that it reads as though one person wrote it.

© 1996 by Holbrook & Kellogg, Inc.

Therefore, this manager makes the writing strategy work.

Staffing is one key concern by which to measure proposal managers. A knowledgeable proposal manager will comprehend that an effective work force has to be productive on the proposal from RFP release roughly up to proposal completion. However, when does the start of the effort occur? How much should this initiation cost?

Figure 7.1 compares manpower *versus* time for proposals. If a firm starts working on a proposal too early, then the burn rate of overhead dollars will be ridiculously high because people do not have even a pre-RFP to read. The three examples shown in Figure 7.1 represent the most common ways in which firms deal with proposals.

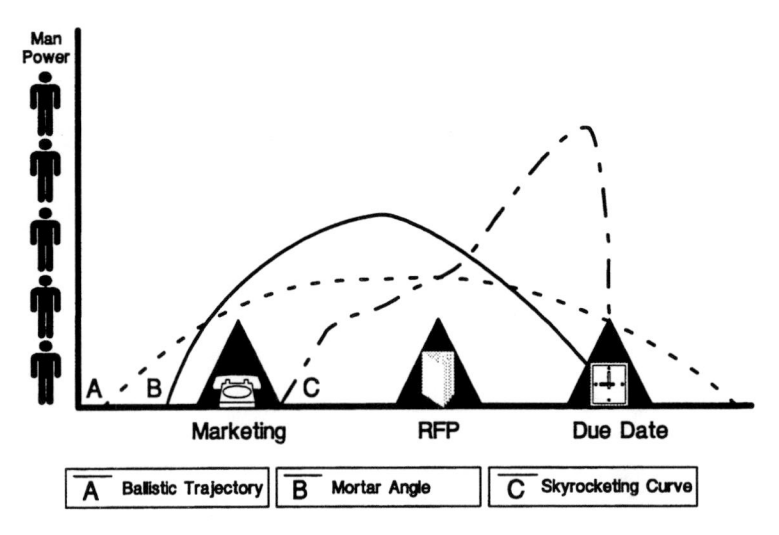

Figure 7.1
Typical Proposal Time Lines

Type A. The "ballistic trajectory" approach in staff loading (Type A) starts slowly, economically, and emphasizes the selection of just the right players; it includes no deadwood and no dead time. This is the best and most effective staff curve. Please note that the downward trend in staffing does not reach zero until well after the proposal has been submitted. Plenty of post-proposal time remains to plan the next steps, the post-submittal marketing efforts. Type A is somewhat ideal—if all proposals were written this way, costs would go down and wins would go up.

Type B. The "mortar angle" approach (Type B) is the industrial norm. This brief angle shoots up once marketing and management anticipate a victory. Activity peaks in the weeks immediately before and after RFP receipt, and peters out rapidly as the due date approaches, while managers try to finish in good order without invoking too much cost.

Type C. The skyrocketing curve in staff loading (Type C) represents every proposal team's nightmare, and is a typical, if not prevailing, experience at most firms once a year. Marketing detects a likely target late in the cycle, management goes for it regardless of costs, and people are mobilized from all branches of the company. Only a few people understand the RFP at the time a crowd of strangers comes in looking for directions. Performance, quality, and cost are now thrown out the window, and the teams plummet headlong just to finish on time. The due date is made with nothing left to spare. (Firms that do all their proposals this way usually fail to meet the deadlines, thereby making the entire proposal worthless.)

A successful proposal manager is always trying for a smooth build-up, settling for the crisis susceptible mid course, and avoiding true last minute stampedes. The more often this manager and deputy manager have seen these dramas acted out, then the more mature, sophisticated, and effective the proposal will be.

The firm must reserve enough time, energy, and resources for the end of each proposal to continue the marketing effort at the client's office, get ready for the client's technical questions, prepare for best and final evaluations, and have a cadre of talent ready to start a new contract. Thus, the proposal manager bears this heavy burden of coordination. As the primary advocate of the proposal, this manager is its champion, spokesperson, and fundraiser through its lifetime. The vehicle used to notify others within the company of how each important stem will proceed is the proposal directive.

The culprit industry-wide in putting proposal teams into the expensive and risky mortar and skyrocket approaches to manpower loading is, and we repeat, is always, management lateness in the Bid/No-Bid decision. Forward thinking firms know 6-12 month ahead of an RFP that they will bid; there is no lengthy process of trying to figure out a simple yes or no. While the clock ticks on managers who can't decide to bid or not bid, the proposal team is running out of time. When a decision to bid finally arrives, management typically thinks the slack in a tight schedule can be made by asking the proposal team to work nights and weekends. It comes as a

© 1996 by Holbrook & Kellogg, Inc.

surprise that the proposal manager may already have counted in such long hours. Untimely bid decisions will create poor proposal quality by depriving the proposal manager of the vital time needed to pull together a quality team.

7.1 THE PROPOSAL MANAGER'S DIRECTIVES

The proposal directive is a written order, which the proposal manager gives to the team. Several typical directives are transmitted by memo in any well thought out proposal. These memos include the:

- Kick-off meeting directive;

- Win strategy, marketing information, and theme development directive;

- Blue Team directive;

- Red Team (or midterm) corrective action directive;

- "Time to finish" directive; and

- Competition directive.

Kick-Off Meeting Directive

This memorandum informs people that they have been selected for full or partial assignment to the proposal team. This directive names the following staff members:

- The proposal manager;

- The deputy proposal manager;

- Technical volume leaders;

- Business volume leaders;

- Staffing and resume volume leaders;

- Cost volume leaders;

- Proposal development organization personnel, assigned for the duration;

- Red Team manager;

- Blue Team manager; and

- Consultants and subcontractors picked for specialized talents.

Other information given by this memo will consist of:

- The responsibilities of every person;

- The deadline;

- The milestones;

- The quality and quantity of work expected;

- The elements of the win strategy in the writing plan; and

- The storyboard's control and use (also see Chapter 9 for more details).

This first directive thus formally consolidates and shapes the team. Typically, the cost code charge number for this proposal is given to in-house staff members for their time sheets. It should be clear from the start that the proposal manager is the boss.

Win Strategy Marketing Information and Theme Development Directive

This memo usually comes out one or two weeks after kick-off. It explains management and marketing win strategies, as well as what we know of the client's technical preferences.

The team then goes into gear, with the proposal manager at the helm. The team formulates a hypothesis of what will thematically appeal to the client, fits this hypothesis into a concise win strategy, and gains as much inside information about this customer as possible. At a later date, about midpoint, these items will be verified and formalized in another directive.

Blue Team Directive

This announcement prepares the writing team for the Blue Team, a special set of managers and engineers who have sized up the potential competitors. The Blue Team evaluates the competition through the following data:

- Competitors' annual reports;

- Competitors' marketing literature;

- Client's appraisals of the competition (by project);

- Knowledge of the mobilization of other firms; and

- A list of how many other proposals to expect.

The Blue Team analyzes these data for two reasons. First, it informs the White Team of what to expect from the competition. Second, the Blue Team synthesizes all the best qualities of the competitors to create the perfect competitor, who would instantly win the contract.

This hypothetical, ideal competitor is important, because while the writing team poured itself into its work, it probably overlooked how to meet, match, and beat the opponents. The Blue Team creates a competitor whose proposal fully matches the RFP, whose project manager is stupendous, and whose success record overall and on similar projects is enviable. The purpose of this paragon shows us that we are not perfect, but there are things we can do to achieve perfection.

Red Team or Midterm Corrective Action Directive

This memo usually cites some items the proposal manager wants to fix. Red Team findings include:

- Staff additions or deletions;

- Corrective actions;

- New deadlines, extensions to the RFP;

- Amendment information;

- New developments, alterations, or changes;

- Reminders to be security conscious, and shred waste; and

- If used, pre-Red (or Pink) Team findings.

"Time to Finish" Directive

This memo announces the homestretch, and asks tired, worn-out people to hang in there and make the last efforts with all the energy and professionalism they can muster. It is both a pep talk and final push memo.

Completion Directive

What is left? To make sure no loose ends are left, the team must perform certain steps in order to complete their assignments, turn in materials, and stand by for client questions, best and final evaluations, and the award. This last proposal directive also congratulates people for their hard work and accomplishments.

7.2 THE PROPOSAL MANAGER'S DIRECTIVES AND QUALITY CONTROLS

The proposal manager and volume manager ensure the quality of the product by editing it. This step usually is called quality control (QC) but is really editorial in nature insofar as text and art are either acceptable or rejected for quality reasons.

The accepted sections should be praised and the rejected sections returned with suggestions. In the QC step, we weed out text and art that do not fit the writing plan, the RFP, or the win strategy. We also delete text and art that deviate from directives, or that is shallow, imprecise, and unpersuasive. Managers should insist on high quality and coach or replace persons whose work does not measure up. Quality control is an indispensable step in achieving the goals of a successful proposal.

7.3 AUTHORITY AND BUDGET

Some final words are needed to affirm the role of the proposal manager in seeing the product through to completion. Every firm has its own definition of how much authority a proposal manager is given, both in terms of supervising other employees and in spending overhead dollars. The more successful firms vest their proposal managers with ample people and money to get a quality document done on time.

Traditionally, the proposal manager is that single person at the firm who understands the client's problem optimally, and can show others how and what to do, customized and tailored to what the client expects and recognizes as excellence. If a person is selected who is not exactly true to the preceding description, then the staff will have to support him or her at even more dedicated levels than an ordinary proposal. Also, the proposal manager will have to "step up" to the new role—the client's expectations for a single harmonic thought stream will not be relaxed just because the proposal manager has, personally, only 75 percent of the credentials needed.

In such cases, the proposal manager must use the full range of tools available, not the least of which are covered in this book, but the remaining tools must be thought of as people, often one skilled person for every unique or overspecialized area in the RFP. The proposal manager will integrate these individuals into his or her thought processes and into his or her own expertise, yielding a seamless synthesis of narrative and art to show a superior approach to the RFP. Nothing else will suffice—the client can tell if the technical contributors are not in sync with the overall proposal.

The proposal manager has to personally resolve disagreements which can prove to be chaotic to the team, and to assume personal responsibility for the end product. Make sure your proposal managers are invested with enough budget and authority to make the formula work—remember that every winning proposal is completely responsive to the RFP in creative and innovative ways, sells your firm's expertise in a dignified, technical framework, and flawlessly articulates your discriminators and sales messages. Unless the proposal has the time, people, money, and energy to achieve these goals, you may be looking at a no-bid or unintentionally being a runner-up in a race where there can be only one winner.

In closing, 1994 studies of proposal managers around the Washington, D.C. area demonstrated that firms with a high win ratio, that is in the overall range of 25 to 45 percent, all gave their proposal managers these levels of authority:

- The ability to spend 100 percent to 125 percent of the proposal budget, unimpeded.

- The authority to front-end load the proposal, well ahead of RFP release date, staffed with sufficient subject matter experts to excel at expected RFP criteria.

- The authority to shut-down or "freeze" concepts or designs in order to keep the proposal in sync with the proposal manager's master plan. This authority includes the power to cut off engineers and technical people from further developing their ideas beyond where the proposal manager recognizes that benefits are accruing to his goals.

- The authority to review, affirm, deny, use, or modify Red Team results.

Thus, in the end, the proposal is the work of a single person, especially well qualified and dedicated to a unique facet of the client's work.

CHAPTER 8

The Proposal Manager's Tools

Our team is now assembled and comprised of experts in the technology, and a writing plan is well under way. It is time to turn the team to one of the proposal manager's most important jobs—defining a win strategy based on demonstrable facts that a source selection board can rate as technically superior.

Too often, management and marketing people will sift through a small mountain of data and announce, "We are the best!" Unfortunately, at the source selection board, all firms submitting proposals are saying the same thing. Who is the best? The board has the right to compare different firms and judge who is the technically best choice. Proposal managers have to project that superiority.

Information gleaned by management and marketing must be shaped into winning, persuasive technical arguments. Here are a few opinions, not based on fact, that proposal managers have heard:

(1) "At the golf course (or bar), my friend said we could win."

(2) "XYZ firm won't bid because it is doing two other large proposals."

(3) "If we could get John Doe as our project manager, we could win easily."

(4) "Our San Antonio project is just like this one."

© 1996 by Holbrook & Kellogg, Inc.

(5) "We already have two of the 25 centers where the Army does this; we are a shoe-in for this one."

What these people are saying in effect is, "I want to win this one." What they should be saying is, "I believe we stand a good chance to win here." If the company affirms that opinion by undertaking the methodical bid-no bid process, then it has received enough data. In order to supply our company with these data, we must translate our privately held opinions into formal arguments that will technically persuade the customer of our company's superiority.

We must carefully analyze these five opinions in order to prompt questions, which in turn help us to formulate a win strategy.

Number One

On April 22nd, at a business development buffet at the country club of which the company is a member, John Doe told Frank Smith of our marketing staff that the client politely told XY and YZ firms that their performance was unsatisfactory, and thus they should save their bid and proposal budget for later this year when their performance may improve. Because XY and YZ are having such a tough time getting even fair award fees on current client work, we are the third and only untarnished firm thought competent by the client.

We need to know the following: can we get confirmation of the data from two or three other insiders? Are all the data true? Can XY and XZ team with somebody and be perceived as worthy? What mistakes did XY and XZ make in terms of schedule, cost, and performance?

News of XZ's "no bid" on this RFP was leaked to us by a vendor that services us and the XZ company. The reason cited is that XZ is overloaded on two full-scale storyboard and scenario proposals, and it could not find an adequate team to sustain the costs of a third large effort.

Again, we need to know the following: can we get confirmation of this decision from two or three other people, preferably at the XZ company? Would they consider us a likely teammate, and why or why not? Will the vendor's management affirm the sales representative's motivation for seeking us as the potential winner (and source of new orders)? Is this vendor who leaked the information a renegade, or an ally of XZ, YZ, or others?

Number Two

On February 11, 1994, the client's program management office (PMO) General Manager told three of our managers he admired John Doe of ABC corporation as an ideal project manager. He stated this preference at a meeting where we gave our white papers on how well we would do this contract. This same high ranking official told us he would be the authority on the source selection board. He was lukewarm about other people in our firm whom we nominated as potential project managers. Nevertheless, he rated our technical and management approach as the best he had seen.

We need to know: if this client manager is serious, can we get Doe to come over to our side? Is Doe qualified, and would he be agreeable to the board's membership? Is this PMO Manager getting ready to retire or be transferred out for some reason? Is he respected by his own technical staff? Does he have an axe to grind with any of our competitors? Will he have an axe to grind with any of us if we do not hire Doe?

Number Three

The project manager, contracts manager, and engineering manager of our San Antonio office received copies of the pre-RFP and RFP. They see about a 70 percent similarity, with some significant differences in key areas.

Our managers need to know whether there are itemized distinctions between the contract's scope of work and this RFP. As a pre-Blue Team exercise, can they think of anyone who could put together a team and win? What can the San Antonio people remember of their own win strategy, themes, and factors, which the client judged superior? What do our contract files tell us about this client?

Number Four

Although our firm has contracts at two of the 25 centers operated by the Army, each center is drastically different. Also, the 23 other centers have some 23 prime contractors and 35 subcontractors who could make the same claims we do. Many centers are having performance problems, and their contractors have been discredited.

We need to know: has our good performance been recognized in this pack of troublesome contractors? Are our centers perceived as quality conscious? Of the 23 other broadly capable contractors, which team could come across as competent for this RFP?

Number Five

Three weeks ago, the client's engineering staff was briefed by our engineering staff on our white papers and technical approach. Our four engineering managers came away with distinctly different accounts of what the three client engineering managers said.

One of our people, whose area was praised, believes the client wants only our firm. The other members have mixed feelings. One feels quite good about our being selected as the RFP's winner, but only after we make some revisions. The remaining two managing engineers felt that the client was lukewarm to their parts of the briefing, and think major reworking would make our concepts acceptable—acceptable, but superior?

We need to know whether we have an engineering perception problem concerning this client. What posture can we take to win over this client's engineers?

Therefore, we have all the hard, factual data. The team can start to shape its win strategy based on precise knowledge of this part of the marketplace.

The role of the proposal manager now becomes that of a coach. How can the disadvantages we have be turned around? How can we wrestle advantages away from the competition? In the shop talk of proposal development organizations, these factors are broken down into four factors:

- Positive marketing intelligence (the Ah-Ha Principle);

- Negative marketing intelligence (the Oh-Oh Principle);

- Faults of the competition (Ghost Stories); and

© 1996 by Holbrook & Kellogg, Inc.

- Distinctions (the Discriminator Principle).

8.1 WRITING STRATEGIES

From the scenarios developed above, our knowledge base now has some tabulated *Ah-Has,* which management, marketing, and the proposal team all agree are positive. What do we mean by ah-ha knowledge? Let us continue with the previous scenario.

Our two most dreaded competitors, XY and YZ, have diluted their strength elsewhere and at best could only show up as someone's subcontractor, with little about which to boast.

- *Ah-Ha:* "We are the recognized industry leader in this technology (the other two competitors tarnished themselves)."

- *Ah-Ha:* "We are ranked first by the government in this technology (the other two firms were flawed and subject to criticism)."

- *Ah-Ha:* "We have demonstrated management superiority in this project manager (whom the client recommended to us specifically for this contract)."

- *Ah-Ha:* "We have 200 percent in-depth coverage in personnel for the position."

- *Ah-Ha:* "Our Alpha-Omega contract provides us with a testing ground to prove 70 percent of this scope of work."

8.2 THE OH-OH PRINCIPLE (NEGATIVE INFORMATION)

The proposal team is then assigned to explore and subdivide these Ah-Has further into thematic sentences for major heading in each volume. They experiment and test each for impact, provability, and consistency.

The exercise of strategy now goes on to other factors that may be just as useful as Ah-Has. We do have some things to protect, disguise, and shelter from abusive remarks by other firms' proposal terms.

- *Oh-Oh:* The client's engineering staff is not impressed by our engineering concept. If we claim unbridled technical excellence, someone is likely to cut us off—at the knees. We may even have one or two outright "no" votes at this point from people who could very well be on or advise the source selection board.

- *Oh-Oh:* We operate only two of 25 centers like this contractor; many firms out there can say, "Me too!"

- *Oh-Oh:* The proposed team has tallied up that they have only six of the needed 12 staff managers to claim 100 percent experienced staff managers.

- *Oh-Oh:* Several smaller competitors could team with XY and YZ, call this a perfect match, and underbid us.

- *Oh-Oh:* Five years ago, we had a disagreement with this client and made some enemies. Some have retired, some were transferred, and a few remain. Those who remain have told our competitors we have flaws.

The proposal manager now has to make trade-offs in order to emphasize team strengths and compensate for shortcomings. The mark of a good proposal manager is that he or she faces up to bad news and copes with it. An immature proposal manager will take the politically expedient way and deny the company has any problems. This is the so-called "perfect company, perfect sucker" approach.

In the course of our intense scrutiny of competitors, we discover that some of them have telling faults.

8.3 COMPETITORS HAVE FAULTS, TOO (THE GHOST STORY PRINCIPLE)

Our examination of the other potential teams has yielded some pretty impressive results. Of the 20 or so firms employed elsewhere by the client, in somewhat similar endeavors, none has the facilities, backers, or bankroll to sponsor this contract with any credibility. In fact, as we suspected, only XY, YZ, and an odd assortment of runner-ups have any hope of competing in light of the advance marketing done by our team.

© 1996 by Holbrook & Kellogg, Inc.

The news gets even better, for not only are XY and YZ hard pressed to make a showing here, but they have had some corporate black eyes administered recently by clients:

- XY's award fee on its contract from this client was in the 60 to 70 percent range: a *Ghost story* on poor or marginal performance.

- XY's project managers have been criticized by the client, and a few have been replaced: a *Ghost story* on weak management.

- XY teamed with a number of smaller firms and was beaten five out of six times: a *Ghost story* on weak, implausible teams.

- YZ has made enemies in several states, where the governors and legislature find YZ's name the equivalent to political treachery, double-dealing, and poor community relations: a *Ghost story* on nonperformance.

- YZ's managers are politically motivated and market rather than attend to technical business; they are in ill repute at the client's PMO: another nonperformance *Ghost story*.

- XY and YZ have well over the industry's turnover rate in white and blue collar labor categories: *Ghost stories* on poor personnel practices.

Our business development people have uncovered other unflattering items about the chief competitors and allies. We decide to use some of these data as *verbal marketing data,* articulated *in person* to the client at lunch, et cetera. The remainder of the data goes into the proposal. This strategy protects our weaknesses, hits theirs hard, and gets every ounce of strength out of our own track record.

The proposal team is poised to synthesize all the Ah-Ha and Oh-Oh information and to evolve the data into Discriminators that will push through the opponents' arguments and win.

8.4 DISCRIMINATORS

The compilation of all this verified information is the *Discriminator,* which is an item to which an opponent cannot respond with a reasonable answer.

Like in a debate, we try to catch the opponent flat-footed, off-balance, and without a reply. When evaluators read the sections of the opponents proposal *versus* ours they must see these differences, these Discriminators. The proposal manager and his or her team have to include discriminators in the proposal, as we show in Table 8.1.

To show we know the difference, our proposal manager assigns the Discriminators to volume leaders who define and make actual, embedded sales messages fit conspicuously into the text, such as "with an industry leading rate of only seven percent attrition in programmers, engineers, and managers, we offer the most stable, reliable personnel base for the contract's lengthy 60-month period of performance."

Adverse Iteration

Adverse iteration converts possible negatives into definite positives. Adverse iteration is the art of taking a negative definition or situation and turning around the material to show positive elements, such as:

- *Negative:* Our engineering staff had three of seven tests fail.

- *Positive:* The majority of tests run by our engineering staff succeeded.

Here is another example of an adverse iteration: our engineers planned and managed four of seven tests completely successfully; that is, 57 percent of the first run tests went flawlessly. On the remaining three tests where flaws were encountered, complete *reliability availability maintainability* (RAM) calculations were performed to test the faults out of the system.

- *Negative:* The client does not like some of our engineers.

- *Positive:* Many of our engineers have excellent track records.

Table 8.1
Using Discriminators

Ah-Ha, Oh-Oh	Ghosts	Can They Reply?	Proposal Sections	Our Discriminators
XY, YZ have to team.	They are too weak to perform alone.	No	Executive Summary Technical Introduction Business Introduction	We stand alone.
Our project manager is the client's choice.	Their managers are second class.	No	Executive Summary Technical Introduction Business Introduction	We manage very, very well.
Our expertise really is unique.	Theirs is too.	Yes	Similar Experience	* Weakness—we need to fix
Our engineering is not well respected.	They are equal to or better than us.	Yes	Technical Volume	* Weakness—we need to fix
We are the recognized leader.	At present, they are not well thought of.	No	Technical Volume	We have technical excellence.
Personnel is our strength.	They have trouble keeping personnel.	No	Business, Technical, and Personnel	We have the most stable base.

* Weak spot identified—needs adverse iteration that is, turning a potentially negative message into a definitely positive message.

Adverse Iteration: Our proposed engineering staff consists of 54 individuals with an average of 15 years experience on similar projects; 58 percent of our engineers have M.S. degrees in engineering, physics, or hard sciences; and 35 percent have the M.B.A. degree—all are ready to start work on the first day.

Ghosting

The proposal manager details and expands these messages to section leaders for further elaboration throughout the proposal so that the text will read as though one person wrote it. Having put out a directive containing how we will highlight our accomplishments and protect our weak spots, the attention goes next to how to exploit the opposition's weaknesses. "Ghosting" is the common term for hitting the opposition where they are least prepared to respond.

- *Their positive claim:* XZ is the only firm skilled in specific engineering skills for this contract.

- *Our negative reply:* Of the many firms capable of performing the work, we offer . . .

- *Our ghost on their performance:* The engineering tasks in this scope of work are unlike our earlier contract in that different types of skills will be needed.

- *Their positive claim:* Our management is excellent and will respond to every part of the scope of work.

- *Our negative reply:* We regularly are awarded a score of 85 to 92 percent of award fee on our contracts. We cannot say we know the competitor markets all the time, so our boast allows the client to check similar experience and find out (a) company XY had only 70 to 75 percent of award fee, and (b) contracting officers have had some audit agency complaints about XY's excessive marketing.

- *Our ghost on their performance:* With Mr. Doe at the head of a respected management team, we will deliver the project on its tight milestones under its strict budget as our 88.5 percent average award fee demonstrates. (While the competition may promise the same thing, we will win because (a) the client handpicked our project manager, Mr. Doe, and (b) we have the track record for high success rates.)

When written well, the ghosts affirm our own strengths. As we write more adverse iterations and ghosts, we get a more thorough view of how strong we are. The proposal manager and volume leaders test each statement to see if they are:

- Credible and easy to prove;

- Old, bizarre, hard to understand, or humorous (these are the worst kind and can cost us points);

- Compatible with the win strategy;

- Not insulting or abrasive to the client;

- Pertinent to the RFP; and

- Persuasive without being argumentative.

Lastly, can we be sure the client will not respond, "Ho hum," "So what?," or "How could anyone say this?"

Now, as the proposal writing team pushes into its first drafts, it is armed with definite Discriminators that show we know how to win in a contest. (As we proceed, we call these factors DAOG—Discriminators, Ah-Has, Oh-Ohs, and Ghosts.)

Typical Discriminators

Here are some typical Discriminators that have proven successful in military hardware and software contracts, and are part of a larger thematic pattern of arguments that flows uniformly from start to finish:

- This product is already in production at our Los Gatos facility.

- Our new perspective to this field gives us an objective, innovative approach.

- Our extensive operations and maintenance knowledge of this equipment ensures total control of life cycle costs.

- Our team, composed of people of all disciplines throughout the industry, is the ideal blend of proven hardware and software experience.

- We exceed the demanding performance requirements by 25 percent through an innovative technical approach.

- The hot-bench tests will be simulated ahead of schedule on our proprietary test-bench software.

- We invested $1.75 million in our own IR&D money to qualify this item in its pre-RFP stages.

- Our superior engineering division is enhanced by our subcontractors' operations research centers.

- Our company developed software architecture configuration tests, which exceed the MIL-STD performance standards by 50 percent.

- Our test plan eliminates unproven RAM elements and ensures reduced risks.

- Our Automatic Data Processing approach is as thorough and effective as competing packages, and functions from a single critical path and a simplified work breakdown structure.

- We have tested this design to 99.9 percent reliability.

- Seventy-five percent of the hardware and 55 percent of the needed software exist in our inventory from similar projects; all that remains is to engineer the interfaces.

- Our automated design tools provide the means to have a preliminary design in 30 days.

Thus, preplanning has given the team a cache of ammunition to use to win, and a set of examples on which to draw.

8.5 THE RED TEAM

An essential ingredient of the proposal and a key tool in the proposal manager's bag of tools is the Red Team. Some firms do not formalize the Red Team, but instead call them management reviews, where top management reads the proposal and RFP skeptically, and redlines the document. Both approaches pay off, but the true Red Team consists of dedicated outsiders who bring a fresh perspective to the writing team and especially to the proposal manager.

The Red Team usually has a "captain" or leader who tallies overall results and formulates criticism into channels where the proposal manager can respond to individual ideas, without lost time. Often, the Red Team captain was the firm's second choice for proposal manager—that is, someone who is exceptionally well qualified to speak on the issues. Red Team members are employees drawn from allied fields, people who know the client and proposals, or consultants with proposal critique skills. They assemble, preferably over a weekend, read the RFP, read the proposal, critique it in positive, affirmative terms, and make recommendations. They are then free to return to their jobs, unless the proposal manager affirms critiques and asks selected persons to stay and make the "fixes."

In combative situations, Red Team captains who genuinely believe the proposal is flawed as-is and who consequently have no faith in the proposal manager and approach, will ask corporate management to overturn—that is, "fire" the proposal manager and make a new appointment, with collateral mass changes. That is the exception that proves the rule. Most proposals begin and end under one manager, who absorbs what he or she can from the Red Team and presses on.

Here are some Red Team factors that the proposal manager will want to have in mind:

- Select members for their management skills.

- Select members for their technical skills.

- Be sure you have no mavericks, rogues, or proverbial "loose cannons."

- Schedule the Red Team for a weekend, or Friday, Saturday, Sunday, Monday as a worst case.

- Don't let consultants overload or overdo Red Teams; critiques have to be useful, pragmatic, and economic.

- Get the best Red Team you can find; try to disarm all the client's and competition's known strengths when you rewrite.

- Itemize critiques by RFP number, result, cost, and impact.

- Accept and revise for as many critique items as you can beneficially use.

- Don't be intimidated by the Red Team—they work for you and are there to help you improve your product.

- Don't get belligerent with the Red Team, even if they become offensive; they perceive that they are doing their jobs.

- There are very rare incidents where proposals die a natural death because events have transpired that make it politically impractical to propose further; don't be discouraged if one proposal out of twenty ends this way, or from lack of funds or even lack of interest.

- Don't ever, ever turn in a proposal without a Red Team.

In brief, the Red Team is one of the proposal manager's most important tools to be used to get a fresh perspective on the RFP and to see the problems through the eyes of other qualified individuals. No proposal is complete or worthy of the client's review without a Red Team.

CHAPTER 9

The Storyboard, STOP and Other Writing Techniques

While the proposal manager pushes ahead full-speed on the actual technological and business aspects of the proposal, he or she must make sure the deputy proposal manager is handling the most important aspect of the proposal development group: the *storyboard*.

The late Howard Hughes is said to be the father of the storyboard. In the late 1940s, he entered the engineering areas at Hughes Aircraft to discover engineers writing complex matters in small, isolated groups. Hughes supposedly said something like, "Why don't you write your aircraft proposals on storyboards, like we do movies, over at the (Hughes) film company?"

Naturally, nobody had an answer. Everyone assumed that if the clients liked all the pieces, they would like the whole product. Some engineering managers edited the entire document part-by-part, but the document was made consistent often at the customer's expense, when a contract and product hardware were already awarded.

Hughes showed his staff the storyboards for various dramas in the works at his movie company, and is said to have taken his staff over to Walt Disney studios to see how animated films were drawn up; artists and technicians creating a movie is analogous to draftsmen and engineers putting together all the parts of an airplane. The messages from Hughes went to heart; good proposals must be laid out like scripts in

© 1996 by Holbrook & Kellogg, Inc.

order to be consistent from start to finish, and better proposals are put up on storyboards where all the connections and themes are evident.

A number of firms do the storyboard exceptionally well. Their methods and models are industry standards and should be copied by firms with smaller numbers of employees. These firms are Hughes Aircraft, Martin Marietta, Boeing, Lockheed, TRW, Vitro Laboratories and others.

Most large firms have proposal development procedures, but a lot of them are uncreative, unclear, and win by accident. Many medium-sized and small firms and private consultants either know what this client wants, or lose many bids and proposals for every one they win. The storyboard, either scaled down to smaller efforts or full-scale for major proposals, is very effective for several reasons.

First, most professional people *write technically* only a small percent of the time. Most managers get their work done by verbally telling their employees what to do; sometimes they explain work in memos. Also, most engineers get paid for their ability to measure things mechanically, electronically, or in some other technical sense; short reports and specifications are their normal products in writing. Quality assurance people write little more than "It does not meet the specification." Technicians hardly ever write anything from scratch; instead, they take notes.

Another reason why storyboards are effective is that most technical people do not brag about how well they write. Because most engineers and managers do not have to write a great deal, they are reluctant to take individual courses in writing. They prefer to attend collective seminars on proposal writing from a recognized authority.

People are nervous the first time they use the storyboard technique because it looks complex and unwieldy. There is a lot to read and a lot of other work to do. Nevertheless, the storyboard effectively renders a consistent proposal document. The proposal manager puts all the staff together in one place, where the teams must interact.

9.1 WRITING TEAM AND STORYBOARD

The team receives its assignments from the proposal manager, and on a predetermined date, hands in a rough first draft. The team members interact with each other, and read each other's work. The first draft is posted on the storyboard around the conference room so that everybody can read and comment on the sections.

The close proximity of the writers and easy accessibility of the rough draft are beneficial. Writers see things they like or dislike on the storyboard and are free to copy, delete, or add items. Omissions of work are evident to everyone. The team is motivated to produce rapidly.

In industry, the storyboard is usually called STOP for *sequential thematic organization of proposal.* The hexagon is a universal symbol of this technique (Figure 9.1 shows the geometric symbols for STOP).

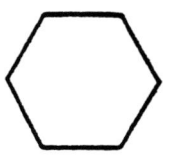

Figure 9.1
STOP Symbol

The storyboard gives the proposal manager an edge on the team members for the following reasons:

- The progress of each individual is posted on the storyboard every day, or at least twice a week;

- Even the crudest field engineers do not want to be seen as ineffective. Therefore, the proposal manager can get everybody to write something; and

- Lateness of and omissions from the document are painfully obvious. Any manager trained in exception management can allocate resources to sore spots.

Figure 9.2 (a, b, c) illustrates typical STOP storyboard sheets handed out to contributors for their work. These may be reproduced and used by our readers. The advantages of using STOP sheets are numerous. First, everyone uses the same format. Second, we can keep track of and manage the assignments easily. Third, writers get used to standardization.

However, we must be cautious of STOP technology in certain respects. For example, we must not start STOP writing until the RFP comes out. Writing based on the pre-RFP may become outdated. In addition, someone from the proposal development organization needs to be on hand to tell contributors what to do, and when. Also, one or two clerical people need to keep the storyboard up-to-date; hard copies of the whole text need to be made for people to study at home or in their offices. (For security purposes, only a few such copies need to be made.) Thus, the storyboard contributes a great deal towards the solidification of the proposal, and provides a visual stimulus to the writers.

WRITING ASSIGNMENT	NAME
	VOLUME
	SECTION

MESSAGE

SALES POINTS AND DAOG

SUPPORTING INFORMATION

ARTS & GRAPHICS IDEAS

Figure 9.2
STOP Storyboard Sheets—Sheet (a)

© 1996 by Holbrook & Kellogg, Inc.

| RFP Para. No. _____ | Title _____ |

Proposal Vol. _____ Section _____ _____

TOPIC: (Write heading that summarizes section contents)

THESIS: (State argument or proposition to be proved in writeup)

MAJOR POINTS: (List salient points supporting thesis)

SCRIBBLE SHEET THESIS METHOD

Figure 9.2
STOP Storyboard Sheets—Sheet (b)

© 1996 by Holbrook & Kellogg, Inc.

Rough Out Art and Tables, Including Captions

Author _____

Figure Title _____   Figure Title _____
Probable Size: ▫¼ pg ▫½ pg ▫full page ▫foldout   Probable Size: ▫¼ pg ▫½ pg ▫full page ▫foldout

Figure Title _____
Probable Size: ▫¼ pg ▫½ pg ▫full page ▫foldout

Approved _____

Figure 9.2
STOP Storyboard Sheets—Sheet (c)

© 1996 by Holbrook & Kellogg, Inc.

As the storyboard matures, the team will arrive at an *exploded table of contents,* which outlines the requirements of the RFP. The team shows off its competence and capabilities in the exploded table of contents. For example, if the RFP specifies that the proposal must contain a quality assurance approach, the team uses all its resources to expand (explore) the criteria demanded by the solicitation (Table 9.1 shows a typical example).

Table 9.1
Quality Assurance Approach—Exploded Version

1.1 Overview, Summary of QA, QC, QE, T&E

1.2 Quality Assurance Provisions
    1.2.1   MIL-STD-9858A Provisions
    1.2.2   In House

1.3 Quality Control Provisions
    1.3.1   QC Techniques
    1.3.2   QC Forms and Instructions

1.4 Quality Evaluation Parameters
    1.4.1   Software Evaluation to DOD STD-2167
    1.4.2   Software Evaluation to DID/CDRL

1.5 Test and Evaluation
    1.5.1   Test Tools
    1.5.2   Evaluation Criteria

If we continue to enumerate the entire RFP in this detail, we obtain a table of contents that articulates our knowledge, skills, and talents, all in a framework still matching the RFP. Only our secondary entries deviate from the RFP, and even so, they are supposed to show what we know. As shown in Figure 9.3, STOP or storyboard technology is logical in the extreme. Sales messages in the form of Oh-Ohs, Ah-Has, Ghosts, and Typical Discriminators occur where the stop signs are shown—where the client's readers are sure to find them. The abstract, or executive summary, precedes the main body, and saturates the reader with the sales messages.

Abstract → Main Body of Text

Main Body
Introduction: Intro | Evidence | Eval

Evidence
Introductory Section: Intro Sect. | Sect. 2 | Sect. 3 | Sect. 4 | Sect. 5

Section 2
Introductory Paragraphs: Intro 1 | 2 | 3 | 4 | 5
Technical Paragraph

Paragraph
Introductory Section: Intro 1 | 2 | 3 | 4 | ■
(Topic) Sentences

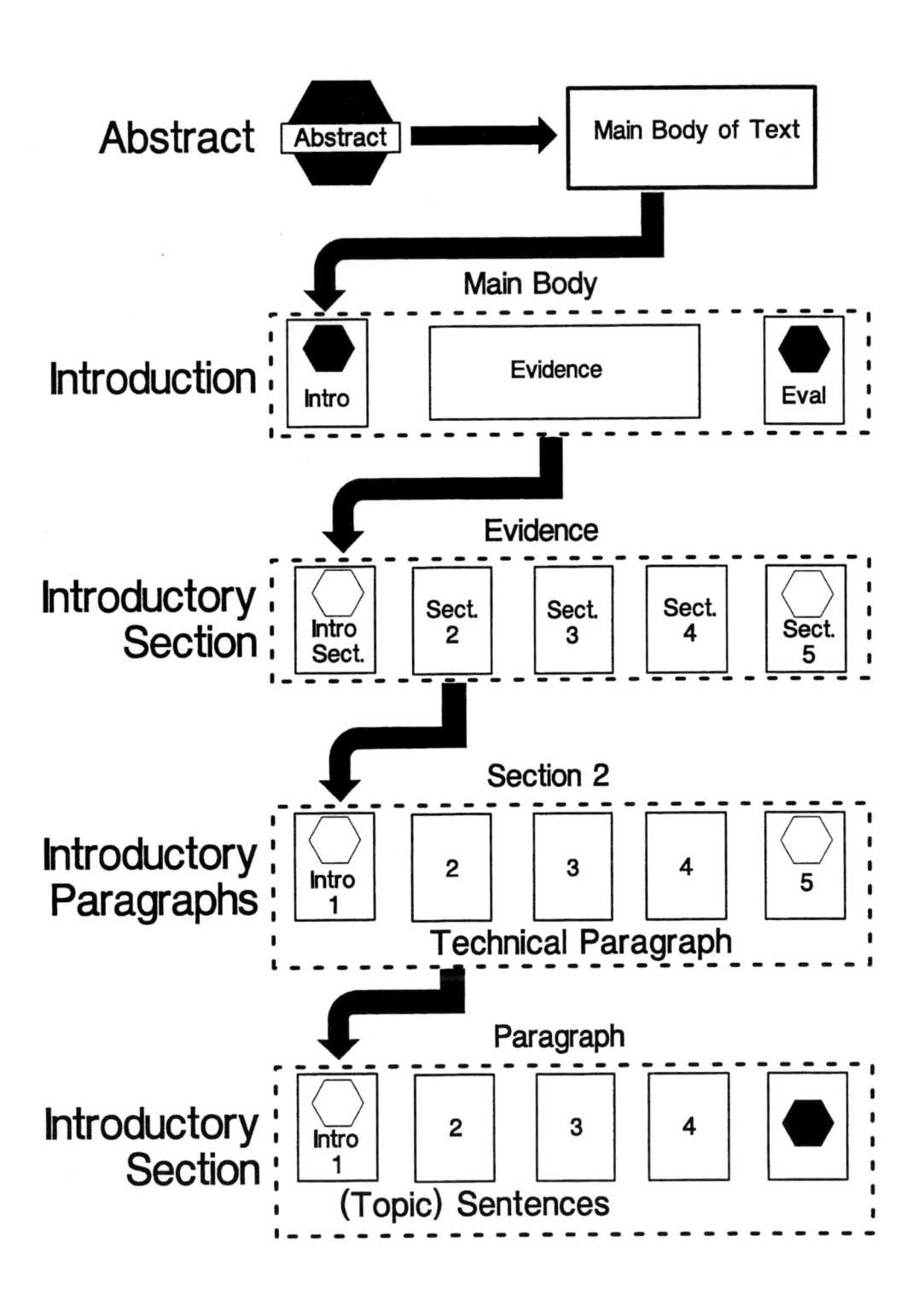

Figure 9.3
"Whole before Parts"

© 1996 by Holbrook & Kellogg, Inc.

As the STOP technique proceeds to second and third drafts, the art and text unite. As the proposal manager and deputy push the STOP format for all it is worth, a product emerges that is thorough, responsive, and professional. Sales messages (the DAOG process) continue to be emphatically placed and reiterated throughout the proposal. Things are starting to pull together.

It is important that managers and engineers on the proposal team insert collective sales messages into critical parts of the proposal. In essence, in a large written work, an element of DAOG should be included in the text and artwork:

●Headings—the major heading titles in full proposal sections (1, 2, et cetera) with message titles;

●Decks—second level headings (2.1, 2.2, 2.3, et cetera) coupled with persuasive titles; and

●Leads—paragraph openers, thematic introductory sentences.

9.2 OTHER WRITING STRATEGIES

The storyboard facilitates the review done by management, marketing, the Blue Team, and the Red Team. The fact that the various players can change the storyboard is very useful. The Blue and Red Teams hit omissions hard. "Why did so much work go into so many blank pages?" they ask.

The ideal storyboard has all the virtues of the Hughes Aircraft method and all the virtues of the Martin Marietta method (both methods are presented in the Appendix).

Management and marketing can see firsthand that their advice and information are being heeded. The Blue and Red Teams can see (and challenge) the product at their intervals, and judge the results of Blue and Red rewriting efforts. Lastly, the proposal manager should "walk the boards" every morning and evening to keep pace with progress, setbacks, and new developments.

9.3 THE HUGHES AIRCRAFT METHOD

In this author's opinion, the classic approach to STOP portrays the managers in a somewhat less authoritarian, more democratic role. In J.R. Tracey's article in IEEE's *Professional Communication Journal*, the Hughes structure works well for people who have worked well together in the past. On the other hand, strangers, who usually are present in a proposal team, may not respond well to a structure that is not strongly centralized. Of course, strong proposal managers make production flow.

9.4 THE MARTIN MARIETTA METHOD

A STOP version by Martin Marietta-Orlando Aerospace is concept-centered on management that yields a uniform product, as though it were produced by one mind.

STOP pamphlets have been in circulation for many years, and have done a great deal to make STOP an accepted part of proposal technology. The union of art with text, and of product with themes, is especially well done. New proposal managers will enjoy the flux and flow of work in the Martin Marietta method.

9.5 TYPICAL SHEETS FOR STOP

During the past decade, some companies have developed STOP in terms of STEP, *sequential (or successful) thematic engineered proposals*. Let us compare and contrast these two quite similar approaches by first enumerating different STOP sheets:

- Lists of Discriminators, Ah-Has, Oh-Ohs, and Ghosting;

- Proposal scenario sheets—text;

- Proposal scenario sheets—art;

- Storyboard sheets—text with art; and

- Red Team storyboard review sheet.

9.6 TYPICAL SHEETS FOR STEP

STEP is a test version of STOP, which is out for evaluation to see if it is a step forward, or a parallel motion elaborating on existing ideas. It has some things in common with Boeing's storyboard technique.

Many advocates of STEP feel it is more thorough, and forces the team to go into more detailed, more frequent checks that will yield better results—especially among people who have not worked together before, or people who do not write frequently. STEP, therefore, is committed to a more intensely managed, more specific format. The STEP sheets include:

- Fact sheets, firm versus competitors;

- Competition strength sheets;

- Competition identification sheets;

- Qualification sheets—our similar experience;

- Qualification sheets—their similar experience;

- Pre-RFP conditioning sheet;

- Subcontractor evaluation;

- Win plan;

- Strategies and headlines;

- Variation differential (us versus them);

- Storyboard plan; and

- Storyboard sheets.

We should note that each STEP sheet is accompanied by a set of 15 to 25 checklists. However, this book does not address STEP in depth for several reasons. First, STEP is the product of a few military software or hardware firms. Second, STEP needs to be precisely tailored to a firm that has not used it before. Third, STOP, when coupled to demanding proposal directives and scenarios, may roughly equate to STEP's advantages.

9.7 ARTS AND GRAPHICS IN STORYBOARDING

Persuasive, modern proposals depend largely on good artwork to illustrate messages of technical superiority. STOP and STEP agree that a well-illustrated proposal is a winning proposal. Historically, the storyboard came from a Hughes Aircraft Company format consisting of the following:

- A 55 percent mixture of words vs. art—that is, about a 1 to 1 pairing of illustrative matter to textual matter;

- A good number of exploded-view manufacturing drawings for assembly—in other words, the drawing tree for how to assemble the product;

- Many electrical schematics; and

- A number of photographs and other hardware shots.

These patterns evolved out of the 1950s and 1960s styles for proposals in which manufacturing was at the center of most work. In the late 1980s, services and software have taken away a lot of the limelight (and budget). The old rules for STOP are not responsive today.

- The format of having one page of text facing each piece of art is no good today. Service proposals typically have one page of art (or a table) per three to four pages of text.

- Few proposals consist of exploded views, except those of major manufacturers, so many service proposals depend on flow charts and tabular data to break up the monotony of text.

- Because manufacturing is largely in second place behind services, the schematics of hard technology are giving way to flow chart symbols. People on proposal teams should know both IEEE and IBM standard symbols verbatim. (Figure 9.4 shows these useful and eye catching illustrations.)

- Large manufacturing facilities or hardware test fixtures are the best subjects for photographs; less complex subjects are easier to draw than to photograph.

Today, as never before, proposals need good illustrations. However, the tendency in industry is toward services, so it is beneficial for proposal organizations to do the following:

- Keep a file of standard flow charts for business management;

- Keep standard technology flow charts for normal offers;

- Have a current MIL-SPEC/MIL-STD library from which to borrow responsive drawings;

- Read as many subcontractors' and competitors' proposals as possible to get new ideas; and

- Hire creative illustrators to put together new concepts about technology.

Thus, art and graphics are an indispensable part of proposal preparation.

The role of art in a proposal is extraordinary. Poorly illustrated sections of a proposal get little attention because of the monotony of too much text, whereas the pattern of a page of art and a page of text facing each other can be called "window dressing." Most winning proposals have a figure, chart, or table on every third or fourth page. These eye-catching materials allow the reader to relax or even to enjoy the messages.

Uniformity

Uniformity and consistency are other issues affecting proposal art. To be uniform throughout all five or six sections of the document, the artwork has to be drawn by people who work in harmony, and use standard ideas, symbols, and styles.

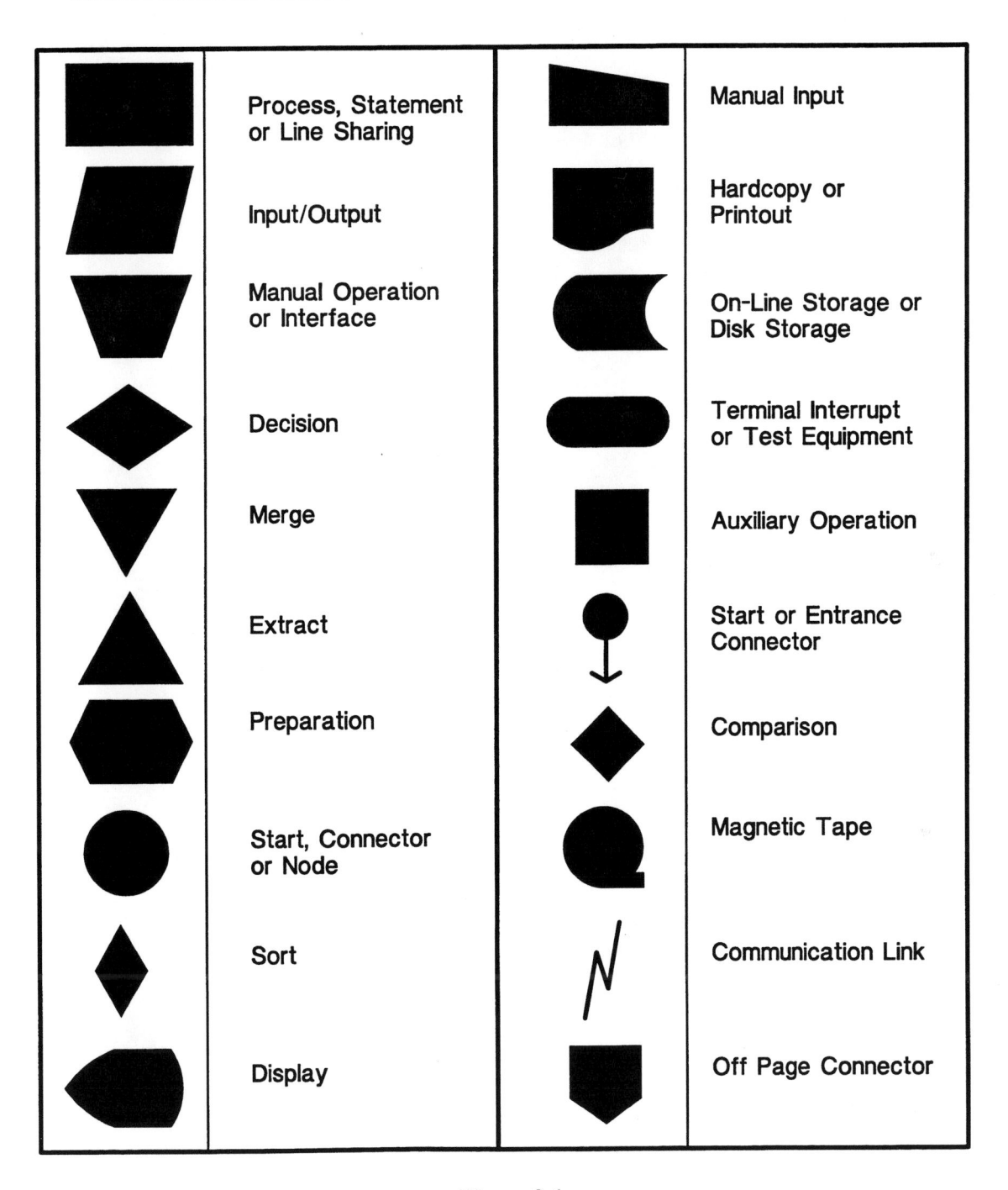

▬	Process, Statement or Line Sharing	▬	Manual Input
▬	Input/Output	▬	Hardcopy or Printout
▽	Manual Operation or Interface	◗	On-Line Storage or Disk Storage
◆	Decision	▬	Terminal Interrupt or Test Equipment
▽	Merge	■	Auxiliary Operation
▲	Extract	○↓	Start or Entrance Connector
⬠	Preparation	◆	Comparison
●	Start, Connector or Node	◓	Magnetic Tape
◆	Sort	∿	Communication Link
⬠	Display	⬟	Off Page Connector

Figure 9.4
IEEE and IBM Standard Symbols

© 1996 by Holbrook & Kellogg, Inc.

Imagine how perplexed the reader would be to see a six-section proposal with six distinctly different types of illustrations! The proposal manager should preview all art in the writing plans, and choose informative, attractive, and standard types of art as models for other work. The storyboard and scenario STOP sheets ought to have blanks just for art, and contributors to the text should each be aware of the placement of art. Rough, handdrawn sketches can be done quickly by illustrators, then finalized after Red Team critiques.

Style

If we wish our artwork to be consistent we must concern ourselves with style, quantity, and placement. *Style* simply means that uniform drawing techniques are applied to all figures. For example, if shaded blocks are used in diagrams and flow charts, then they should be used in all similar artwork. Borders, thickness of lines, standard symbols, type size, and lettering should all be consistent.

Quantity

Consistency of *quantity* implies that if we establish the pattern of an illustration every four pages, then an 80-page chapter will have about 20 figures and tables, which are spaced evenly, and not dumped randomly at the beginning, middle, or end. In the same manner, a section of 20 pages should not have 10 illustrations lest it seem deceptively padded. The text and art must flow together logically. "See Figure 3.1" will not suffice. The drawing must *mean* something to the reader, and be explained in the text.

Placement

Placement of artwork lends consistency. The figure or table should always follow its first mention in the text. Reviewing the art and text together will show the careful reader how to emphasize themes, discriminators, and sales messages. Mundane, unimportant, or lackluster ideas do not deserve illustrations. Summaries, conclusions, overview, and crucial or focal ideas deserve illustration. The placement of a table or figure every third or fourth page should not be automatic; it should spotlight a concept that carries forward the win strategy. The team will achieve winning emphasis when the art and text blend harmoniously, with the sales messages embedded in the words and pictures.

© 1996 by Holbrook & Kellogg, Inc.

As the storyboard fills out, so should the refinements of the illustrations, until at the end of the effort, we achieve a perfect synthesis of graphics and text.

PERT

PERT (Program Evaluation and Review Technique) and Gantt charts are among the most persuasive forms of illustrations because they show simultaneous events in time, and they show the entire event from start to finish. PERT charts are excellent for showing the primary critical path and its subevents. Nothing else articulates so well what certain relationships will accomplish than PERT. The writer is free to impose his or her own logic on the flow, as long as it satisfies the RFP. (See Figure 9.5.)

Gantt

Gantt charts are named after Admiral Gantt, who recognized that shipbuilding requires many synchronous, related tasks. These tasks can be plotted on a linear graph, which thus facilitates our scheduling and understanding of the entire process. This kind of certainty relieves customers of worrying over where their money is going. Together, PERT and Gantt charts demonstrate to the client that we have a firm sense of the big picture, and we know where all the milestones will be. This certainty reassures auditors and friends alike that the proposal is a true, honest, realistic document, not a pipe dream or a bundle of promises. (See Figure 9.6 for a typical Gantt example.)

If a firm manufactures hardware items, these illustrations are paramount:

- Exploded views, and assembly or disassembly shots;

- Facility and equipment photos; and

- Photos of flights or sea trials.

Like text, art has to have a beginning, a middle, and an end. Abbreviations, acronyms, and coded, shaded, or dotted lines should be easy to follow. A figure caption below the figure should qualify the data if they are dense and complex.

An illustration may be based on existing illustrations in the discipline. Military standards and military specifications supply a wealth of artwork. The new DOD-STD-2167 for software development has excellent artwork to go with a life cycle concept. We can extract and tailor these figures, which are already familiar to the clients and audit agencies, to produce an illustration which is:

- True to the client's goals;

- Generated by the client (at some level); and

- Compatible to the RFP.

The proposal manager should check the RFP reference section to be sure the MILSPEC/MIL-STD citations and art comply with what the solicitation required.

Prepoposal Efforts

Management of business development expresses general interest

↓

RFP requested after a brief scope is published in the Commerce Business Daily, or some other source of information.

↓

Preliminary proposal team is identified by categories of expected skills

↓

RFP arrives: Management assigns a Proposal Manager who makes assessments of apparent skills to meet RFP criteria

↓

Proposal Kick-Off Meeting
1. Management underscores determination to win.
2. Propsal manager names contributors.
3. General advice and analyses based on similar past efforts

↓

Proposal Kick-Off Meeting
- Management - potential profits, manpower commitments, risks.
- Engineering - degree of R&D, level of expertise, scheduling problems.
- Quality - QA adherence to specs, and QC and inspection capability.
- Production[1] - ability to comply and available space/personnel.
- Contracts - legal questions, state and local laws, past similarities.
- Accounting/Finance/Purchasing - statements, fiscal data requested.
- Unique or Special Issues - RFP particular, check for consultants.

[1] or Construction

Final Editing, Final Production of Proposals. Overall Team Review and incorporation of all pertinent information

↑

Final Preparations
Item for item crosscheck of proposals to RFP requirements:
fill out all the Certificates, Representations, and Amendments.

↑

Polished Draft Proposals

| Technical Proposal Engineering | Quality |
| Production or Construction | Unique or Special Items |

Cost Proposal Finance/Accounting

Business/Management Management

↑ **Go**

Go/No-Go Meeting

↓ **No**

Send proposed client "NO INTEREST" letter.

Send intent to Bid to proposed client.

↑ **No-Go**

Go/No-Go Meeting → **Go**

Final Approval of Manager with authority to contractually commit company.

↓

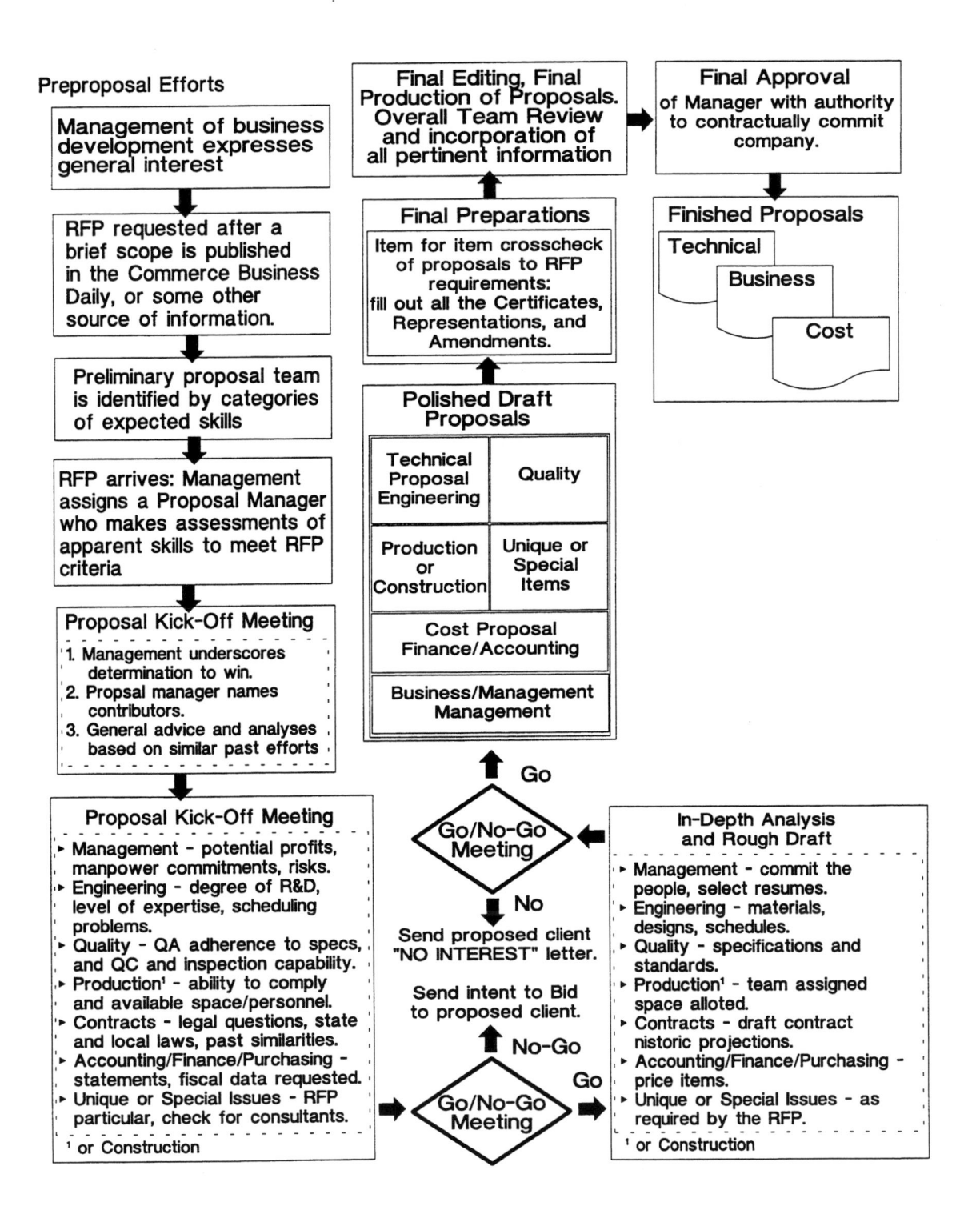

Finished Proposals
- Technical
- Business
- Cost

In-Depth Analysis and Rough Draft
- Management - commit the people, select resumes.
- Engineering - materials, designs, schedules.
- Quality - specifications and standards.
- Production[1] - team assigned space alloted.
- Contracts - draft contract nistoric projections.
- Accounting/Finance/Purchasing - price items.
- Unique or Special Issues - as required by the RFP.

[1] or Construction

Figure 9.5
PERT Chart for Proposals

© 1996 by Holbrook & Kellogg, Inc.

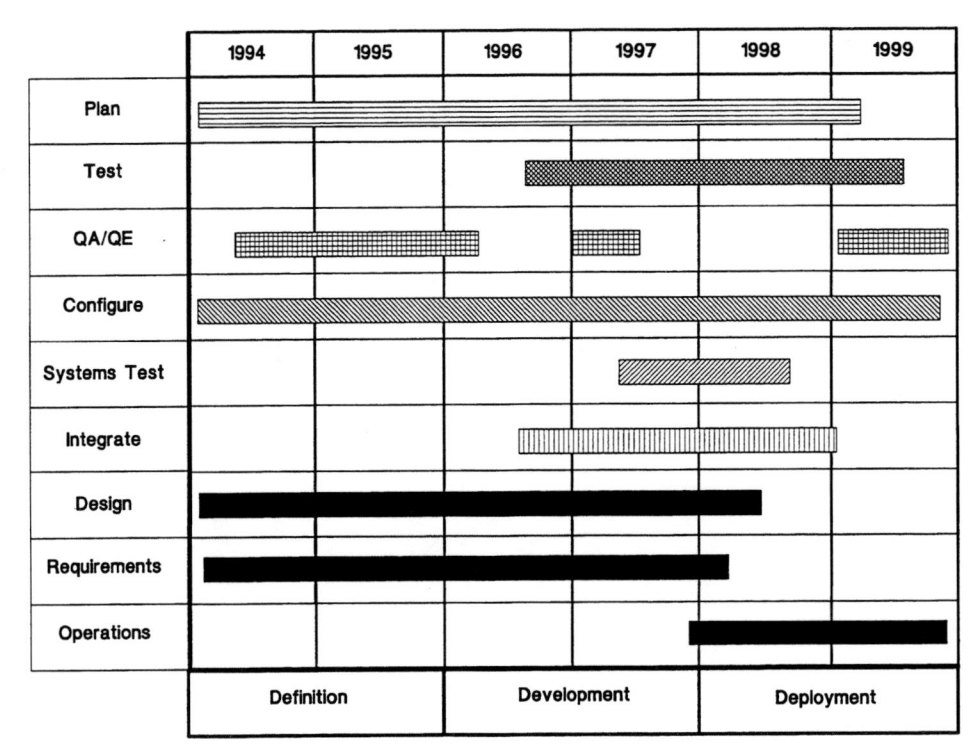

Figure 9.6
Typical Gantt chart for Software Development

By knowing the client well, the proposal manager can make a good guess at how logical figures enhance the argument of a winning proposal.

Because PERT and Gantt charts are excellent for getting to the point, they are excellent for competition. If we compare and contrast the competitors' known methodologies to a plan that is purposefully stronger and tighter, we will produce impressive Discriminators for management and technical performance. When each piece of art supports such a central contention, the effect is overwhelming.

The remarks on art and graphics for this section resulted from a comparison of the mechanical drawing techniques of business, universities, and engineering firms. It is remarkable that since about 1980, the automation of technical information has all but replaced the long-time relationship between the engineer and draftsman. Library shelves of drafting books are now archival only. The illustration capabilities of computers are startling. In fact, many proposal development organizations use artists only for drawing the proposal cover, terrain, or some other aspect in nature which the current computers do not draw well. In short, a proposal shop would do well to be familiar with at least three of the following five software packages, coupled with first-quality laser printers:

- Apple McIntosh McDraw, MacArt

- IBM Graphics

- Wang Drawings

- MacProject—Project Manager

- Harvard Total Project Manager

Thus, with automated drawing packages, we can expedite, standardize, unify, and illuminate the art of technical persuasion.

9.8 USING APPENDICES

By far, one of the most underestimated areas of proposal preparation is the appendix, or a set of appendices, as a part of technical persuasion. In his guide book, *Creating Superior Proposals*, Jim Beveridge states that to keep up the momentum of a proposal, we should only summarize lengthy data in the proposal itself, but save the full version for an appendix. It is true that lengthy chapters and sections slow down the readers, and it is likewise true that such data are appropriate for the appendix.

Let us determine what should go into appendices, and how to select documents for use as appendices. In a specific proposal for Automatic Data Processing services, one common request asks the offerer to show demonstrable evidence in software development management, software quality evaluation, and software configuration management.

The offerer probably already has each of these items in the contract files as documents taken from past and present contracts. These documents are:

- Software management plans in 75 to 100 pages;

- A quality evaluation plan in 150 to 300 pages; and

- A configuration management plan in 200 to 400 pages.

However, we are faced with a dilemma: while excellent, fully developed materials are on hand, they are much too long to insert into the proposal as a chapter or section. To summarize and abstract only the essence of the plans will de-emphasize how thoroughly we, the offerer, know the subject; but these summaries will fulfill the RFP criteria. How can we get the remaining points included for our score? How can we convince software engineers on the source selection board that we are the best selection?

Appendices provide this bridge between size and sense; we can have the lengthy version and the pithy version, and get the maximum score. We must be careful to synopsize the bulky plans down to 10 percent or less of their original size, and still retain quality and all key information. This abstraction of technical data will yield a brief, straightforward proposal section, which concludes with a closing statement, "For a detailed presentation, see Appendix A, Software Plan." The appendix itself must then be sharpened to the new audience specified in the RFP. We must ask ourselves the following questions while reading each appendix:

- How closely does the client's RFP match this precise plan?

- How similar are the old and new clients (if the appendix is a contract deliverable document)?

- What loose strings or peculiar contract clauses are in the plan that could *contradict* other areas of the proposal?

- Are all aspects of the appendix suitable?

- Are standards and specifications called out in the appendix that disagree with this RFP and the philosophy of *this client?*

In the remaining part of this section, we shall pay more attention to the use of appendices in handling lengthy materials, tangential materials, and circumventing page limits.

© 1996 by Holbrook & Kellogg, Inc.

9.9 LENGTHY ITEMS

All too often, an appendix just dumps information into the proposal as though to tell the evaluators, "Here, read this junk when you get the time." In other words, not much time was put into shaping the appendix. This slipshod technique runs the risk of receiving a low score from evaluators, who will be tempted to perceive such an appendix as the product of a lazy proposal team.

The appendix may be crucial to one or two selected evaluators of the source evaluation board, who are, for example, specialists in the discipline of quality assurance. The 25 pages of text we abstract from a lengthy plan has assured them that we are competent and qualified, but so are the proposals of others. Here we use a quality assurance (QA) appendix to explain our superiority once and for all. Based on our latest QA plan, which was already accepted by this client earlier this year, our appendix now completely fulfills the expectations of the evaluation specialists interested in whether our firm can successfully meet the desired quality levels.

However, most deliverable plans are themselves quite long and have appendices of their own—they are not easy reading. They are clearly intended for a select, narrow readership. Like all other proposal materials, the appendices must be tailored and reduced where possible, so that as many proposal evaluators as possible will *browse* through it and absorb our expert treatment. To use a lengthy plan, procedure, or deliverable as an appendix, we should follow these steps:

- Select documents that address the client's immediate needs, or that are similar to those well received by the client; and

- Cut out the superfluous data that obscure the heart of the plan or procedure; the more readable a 100-page document is, the more quickly a skilled or semiskilled evaluator can absorb it. A plan of the aforementioned 100 pages can and should be reduced by one-half.

It takes special skills to trim lengthy documents down to useful appendices, and the company's technical publications department is best geared to edit most of these sections. Other excellent sources of appendices are:

- Feasibility studies done on related matters;

- White papers on RFP related technology; and

- Technical articles by authors who are employees.

9.10 TANGENTS

Appendices are often used to beef up, supplement, or support tangential contentions. If a firm has no true, precise, proven ability to produce a certain level of work, then appendices may be used to enhance the firm's competence. If a firm never had a contract large enough to require separate quality and configuration plans, then a joint quality-configuration deliverable may be used, which will show key elements of both disciplines. The use of tangential appendices may not be enough to win, but it can enhance and elaborate on special knowledge. However, we should use a tangential appendix only when there is no stronger choice available.

9.11 STRETCHING PAGE LIMITATIONS

Since about 1984, many Federal agencies have put page limits on proposals. Despite the dollar value, complexity, or technology involved, government evaluators may insist on 100 to 200 pages in the technical proposal. Fortunately, appendices are often not included in the page limit and are not regulated by the RFP. In such a case, if a 30-page section is cut to five pages, the selectively structured remainder may be suitable as an appendix. Again, the specialized reader will gravitate toward his or her own discipline, and will want to read in-depth data. Therefore, the appendix can be used to circumvent page limits while obeying the RFP.

In Table 9.2, the appendices (in the left column, numbered 3.4 and 4.5) directly follow their related subjects. Thus, their data is more intrinsic to the text and more directly related to what the evaluation must score. If there are a sufficient number of appendices to add uniformly throughout the proposal, then this technique can be very effective. On the other hand, if only one or two appendices are pertinent, they should appear at the end of the proposal, instead of breaking the flow the RFP desires.

Table 9.2 Sample Proposal and Two Detailed Sections	

3.0 Configuration Management
    3.1 CM Theory
    3.2 CM Practices
    3.3 CM Procedures
    3.4 CM Appendix
 	
4.0 Quality Section
    4.1 QA, QE, QC Theory
    4.2 QA Practices and Procedures
    4.3 QE Practices and Procedures
    4.4 QC Practices and Procedures
    4.5 Quality Appendix
	Entire Proposal
	1.0 Engineering Concept
	2.0 Management Section
	3.0 Configuration Management
	4.0 Quality Section
	5.0 ADP Section
	6.0 Similar Experience Section
	7.0 Personnel—Resumes Section
	8.0 Cost Schedule and Cost Control
	10.0 Cost Proposal
	11.0 Appendices

In Table 9.2, however, if 3.0 and 4.0 had page limits, then the appendices would probably push the text over the limit conspicuously, because everybody can see, count, and comment on these extra pages.

We must read the RFP precisely for the use of appendices. If prohibited, the appendices could result in disqualification; if allowed, they enhance the presentation. If the RFP is not clear on the use of an appendix, then we should ask the contact people of the RFP. Placement of an appendix can be an issue.

By segregating data into an appendix at the end, as shown in the right column of Table 9.2, only the interested reader will read it. The appendix's inconspicuous placement at the end is harder for critics to spot. While appendices are sometimes not scored by technical evaluators, they do leave an impression on the evaluator's perception of the proposal's technical competence.

Here are some things an appendix is *not:*

- It is not an addendum—something "added on" because it was accidentally omitted. An appendix is included on purpose for persuasion reasons.

- It is not a dumping ground for leftover proposal materials.

- It is not a place for second-rate, uncorrelated data.

- It is not a step to forgo careful precise management.

Therefore, each paragraph should be examined by the White Team, Blue Team, and Red Team to see where and how appendices can be used to get the upper hand on the competition.

9.12 THE VOLUMES IN SUMMARY

The individual volumes require attention all the way through the proposal process to ensure that we do not omit winning points. Volume managers are sometimes known as volume captains. The cost volume, for example, often lags behind the rest of the proposal for a mere lack of emphasis or interest.

9.12.1 Executive Summary

The executive summary may well be the most important management aspect of the proposal, largely due to the "briefing up" factor. The source selection board reads the proposal and informs the senior agency managers of its findings. The senior agency managers will read only the executive summaries of a few key competitors, and wait to be briefed as to who should win. Therefore, many decision makers will see no more of the proposal than the executive summary. Thus, we come to some conclusions about this volume:

- It had better be persuasion at its best.

- All the favorable political ramifications need to be present or implied.

- The best technical and business reasons should be well articulated in the management, technical, and cost proposals.

- The covers and other artwork must be attractive to entice readers to pick up and read through the executive summary.

- The advantages, including Discriminators, Ah-Has, Oh-Ohs, and Ghosts need to be explicit.

In a typical situation, the chairman of an SSB presents the winner's executive summaries with a large volume of evaluation materials and asks for concurrence. Several things can happen. The chairman may get a signature immediately because everyone is satisfied with the selected contractor's proposal. If the senior agency manager has reservations about the proposal, then he or she may tell the SSB chairman to present all data—that is, to "brief up" all the substantiating opinions and Congressional recommendations.

The senior agency official may reject all the SSB's findings on the following grounds:

- The SSB made a poor selection and should try again;

- The process does not meet Federal Acquisition Regulation (FAR) criteria; or

- Political groups are pressuring for more technically substantial data to offset criticism.

Without a high-caliber, attractive, thorough executive summary, the SSB official is handicapped in briefing up. Suppose that an excellent, winning technical proposal does not have a strong executive summary. The SSB gets the impression that this proposal is engineer-friendly, but not manager-friendly. To compensate, the proposal manager will hurriedly have to prepare materials that other competitors have already submitted. In this manner, technical victors become strategic losers.

The proposal manager should designate an especially responsible, persuasive person, such as the deputy proposal manager, to be in charge of the executive summary. Therefore, SSB officials can "brief up" the attributes of this proposal easily. Under no circumstances should this executive summary be left for the last minute, because it is a key part of success for the technical, business, and cost proposals.

9.12.2 The Technical Proposal

The technical volume must have an excellent foreword and introduction. If the RFP allows, then we can incorporate executive summary materials into the technical volume. The technical volume must reflect all our prewriting and writing strategies, which should display the following:

- Sufficient prewriting research;

- A comprehensive writing plan, which is responsive to the RFP;

- Discriminators in tune with the win strategy and response matrix;

- Storyboard techniques;

- Excellent art and graphics;

- Excellent appendices, if applicable; and

- Good coordination among the White, Blue, Red, and Gray teams.

9.12.3 The Management Proposal

The management (or business) proposal is just as important as the technical volume. The management volume is especially crucial in larger programs. It should stress:

- The credentials of the project manager;

- The credentials of the business manager; and

- The benefits derived from the firm's management structure.

It is a harsh but true statement that firms have lost proposals due to weak management structures. That is, the firm had:

- No credible project manager;

- No credible subordinate managers;

- Commonplace solutions to unique problems; and

- No apparent knowledge of how to manage according to the client's rules.

The proposal manager must be sure that a responsible manager is supervising this section, because without fine management, the proposal runs a considerable risk of failure. Engineers do not always succeed as managers. For a complicated proposal effort, a business administration professional should be included on the writing team. This person should organize the business structure as though this project were already won.

9.12.4 The Cost Proposal

This section is especially significant for firms writing proposals after 1987, because the Gramm, Hollings, and Rudman cost constraints mean a great deal. The cost proposal has to be allowable per FAR criteria, conducive to a work breakdown structure, reasonable, and prudent.

Some proposal writers underestimate the importance of the SSB members who evaluate costs. On the contrary, the SSB cost members become increasingly influential as the competition progresses. They can detect:

- General ineptitude;

- Poor understanding of the problem (from a cost perspective); and

- Bad or sloppy coding for the work breakdown structure, which may cover up a sloppy technical approach.

Cost is an area where exact structuring pays off. Each step and expense must be explained. Consequently, a reliable senior manager or senior cost engineer should be made responsible for every facet of the cost proposal. The proposal manager should devise and supervise the overall strategy so that:

- Our firm will be a qualified low bidder, but we will not bid ridiculously low, and we will not run the risk of being unable to perform;

- We can defend and explain every cost; and

- We can cut a few costs for best and final evaluations to beat others who are trying to win with the low-dollar approach.

Winners typically not only know what the RFP scope of work stated as mandatory, but also the full ramifications of the FAR, MIL-STD, and MIL-SPEC material, item for item.

If a pre-RFP has been released, then we should study the specifications section for the availability and compliance of FAR, MIL-STD, and MIL-SPEC. If we do not have the technical data on hand, we must order them. When the RFP comes out, we should check the reference section for new items.

Although there is a correlation between the qualified low bidder and winner, victory depends on quality, not cost. The *request for quotation* is the vehicle by which costs are scrutinized, not the RFP. RFPs search for a quality consciousness that price alone cannot express. Nonetheless, it is essential not to be perceived as overpriced. To avoid this effect, cost engineers should attend to cost as meticulously as other engineers monitor the technical proposal.

9.12.5 Scopes of Work, Specifications, and Sample Work

Since 1992, the Department of Defense, and especially the Army and Navy, have put a great deal of emphasis on sample tasks and compliance with their scopes of work and military specifications.

When a client requests *sample work,* it means "Let us see you work on this problem as hard and as well as you can."

In 1992, sample work tasks made up about 40 to 45 percent of technical scores in proposals where they were used. Thus, sample work is instrumental in selling good, substantial proposals.

9.12.6 Tailoring Resumes for Proposals

Like other proposal documents, resumes should be tailored to each client and each request for proposal. All too often, resumes are not kept up-to-date and do not portray the personnel as an ideal team.

Out-of-date resumes are the most frequent offenders at most firms. When did our firm last ask every technical and management employee to specify his or her accomplishments, education, advances, and deliverable products for the previous 12 months?

Because management needs to know what the company is doing, it should produce a written resume format. Each professional person should be responsible for articulating achievements, courses taken, and new skills learned. Therefore, updating will be the mere collection of data. Formats are another recurrent problem. Ornate, rigid formats often prevent many employees from looking good. Choose formats that lend themselves to persuasive, attractive layouts, which can be readily tailored to a specific RFP. (See Figure 9.7 for one such resume, geared to the oil and gas industry.)

In order to tailor resumes, the proposal manager must have a good understanding of who will do what, who will be high in the management structure, and what the client's priorities are. A resume should highlight the qualities that make this person right for this job. Tailoring should emphasize work experience, credentials, and organizational know-how. These data may be emphasized in the resumes by drawing boxes around the summary statement, or by citing items in bulleted lists. Lastly, all the resumes should complement each other so that the employees appear to be a *team,* not just a bunch of interviewees trying hard to out-talk each other.

JOHN DOE
Senior Petroleum Engineer
Large Oil and Gas Firm (LOGF)
Tulsa, OK

CAREER HIGHLIGHTS

- *Use this block to highlight and identify special skills and experience related exactly to this client's scope of work.*
- *Refer to deliverables, plans, and hard accomplishments that were this individual's responsibility.*
- *Cite performance-related work efforts, chronologically.*

EXPERIENCE

Mr. Doe is the proposed Manager of Petroleum Engineering for the XYZ Program for the ABC Production Project. He has over 10 years of expertise in oil and gas fields in the areas of production, well sinking, lining, and connecting pipelines.

Mr. Doe's current assignment is in the Gulf of Arabia on a major field development job at a site which promises to yield 6 million barrels per day within the next few months, plus two million barrels per day of gas flow. He supervises a team of six petroleum engineers, two geologists, seven draftsmen, seven technicians, and a field staff of some 50 locals and company personnel engaged in actual drilling operations.

As a Senior Petroleum Engineer at LOGF, Mr. Doe is responsible for overall operations at the site. He administers, negotiates, and manages all properties, equipment, and people at this location, at $4 million per year contract obligation. After his present assignment winds down in three to five months, he will be made available to head the XYZ Program, which is almost entirely similar to his Gulf of Arabia program.

Prior similar work includes Senior Petroleum Engineer and District Manager of the following sites: Indonesian sweet crude project, 1983-87; Gas Development Project, 1988; Abu Dhabi and Saudi Arabian Petroleum Development Venture, 1989-present.

Prior to coming to LOGF, Mr. Doe was responsible for general engineering and property management, reservoir and reserve studies for the Minor US Oil and Gas Co., Ltd. He is an expert in foreign and domestic properties and leases. Mr. Doe has also done extensive logging, design, initiation, installation, and operation of secondary recovery projects.

EDUCATION

B.S. Degree in Petroleum Engineering, 1983, University of Tulsa.

M.S. Degree in Petroleum Engineering, 1986, University of Tulsa.

Advanced Petroleum Reservoir Engineering Studies from the University of Oklahoma, 1990. Registered Professional Engineer in Oklahoma, Texas, California, and Louisiana. Member: AIME and AAPG. Four articles in *Oil and Gas Journal*.

Figure 9.7 *Balanced Resume*

© 1996 by Holbrook & Kellogg, Inc.

Lastly, we show an idealized proposal room in Figure 9.8, a spacious, secure area with ample room for hanging the storyboard sheets around the walls not only in the conference room but also along the walls of the contributors' workstations. The physical space of the proposal room is important because the proposal room becomes the focal point for visualizing the proposal as it develops and is visually shared among the writers through the communicative physical act of hanging up the storyboard sheets where everyone can see them. The dedicated space underscores a commitment to working just on the proposal -- the actual posting of the storyboards, then drafts, then final text are another psychological tool of the proposal manager to assure that peer pressure is brought into play with perennial late contributors. If your material is late, it's not posted to the boards, and blanks go up to show you're overdue. Not even the thickest skinned engineer or manager wants to see his or her work highlighted in this manner, especially if that work is late, sketchy, imprecise or not up to the standards of the other writers. In this manner, the physical presence of constantly updated storyboards is a device of powerful peer pressure, one more STOP tool in the proposal manager's kit, another method to get authors to excel.

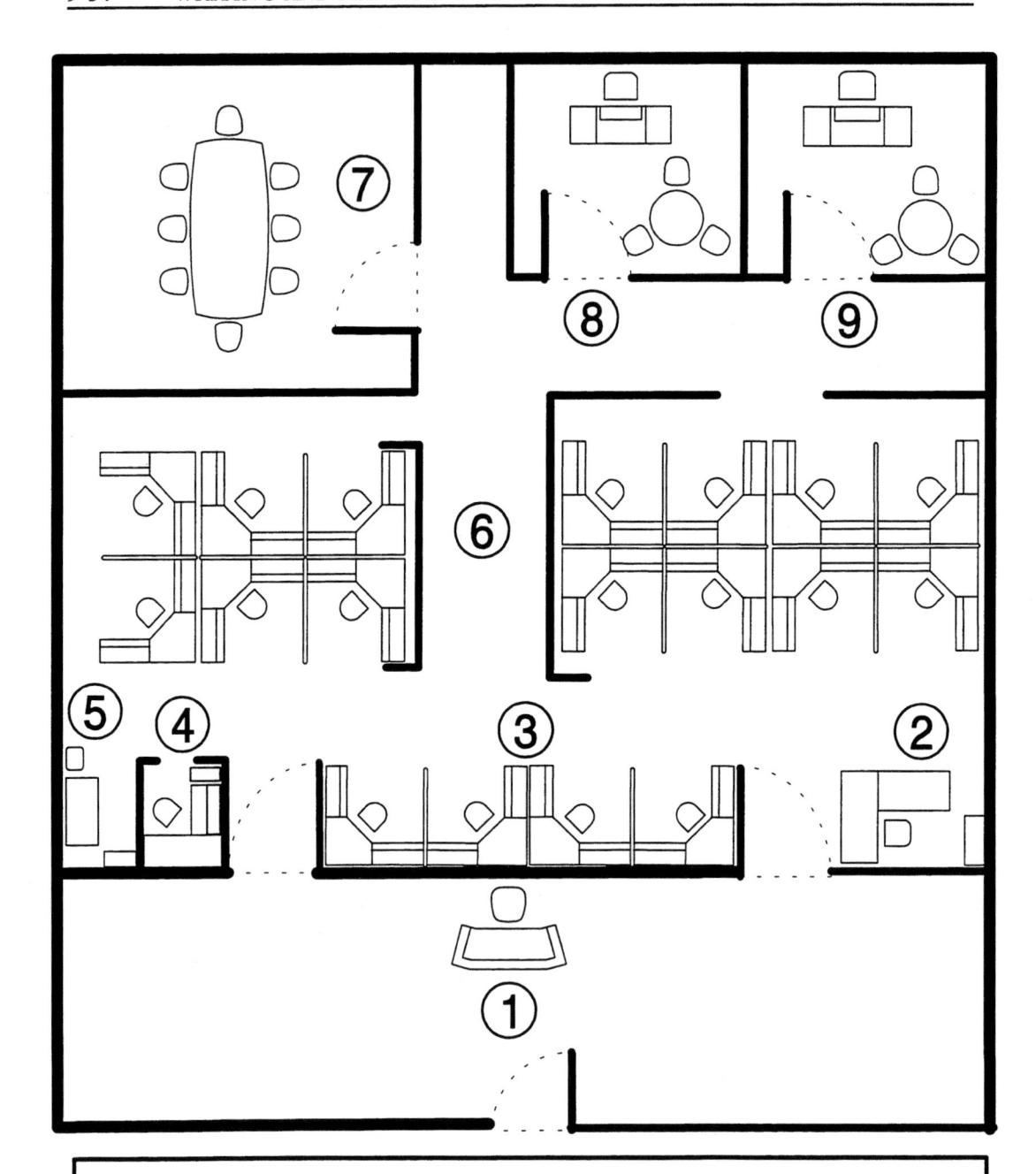

1. Receptionist or Guard
3. Word Processing/Graphic Workstations
5. Copy Machine/Shredder/Supply Area
7. Conference Room
9. Deputy Manager's Office
2. Office Manager's Workstation
4. LAN/System Administrator
6. Contributors' Workstations
8. Proposal Manager's Office

Figure 9.8
Ideal Proposal Room

© 1996 by Holbrook & Kellogg, Inc.

CHAPTER 10

The Final Product[*]

10.1 STOP WRITING

This discussion of *sequential thematic organization of publications* (STOP) format is intended for scientists, engineers, and marketing specialists, not technical writers. However, many points in this discussion are useful for technical writers, who may have excellent communication skills but could profit from a review of the principles and value of the STOP format. Because one goal of proposal writing is to have the entire proposal read as though it were written by one person, the entire proposal team should be familiar with STOP format.

In STOP format, a proposal or report is organized into two-page or four-page units. The main idea of each unit is stated at the top of the left-hand page, and is developed by the text on the left-hand page and the illustrations on the facing page. (In loose STOP format, there may be some art on the left page and text on the right.) STOP units are only two pages long or, at most, four, so that each unit is a manageable size and focuses on the main idea—the thesis—of the unit.

Developing the Thesis

Developing a clear, concise thesis for the STOP unit is a very important feature of STOP format. The thesis of each STOP unit should clearly state the win themes

---

[*] *Chapter 10 was contributed by Dr. Donald Samson of Radford University in Radford, Virginia. Dr. Samson has a distinguished background editing technical proposals for Martin Marietta's Orlando Operations office.*

© 1996 by Holbrook & Kellogg, Inc.

supported by the unit. For example, in a STOP unit on test programs for a weapon system contract, the STOP unit thesis might be: "Our success in testing weapon system components will ensure an effective system safety program." In the unit, main points about the company's testing experience support the thesis. Each supporting argument (three or four to a two-page unit) would be developed with specific details about testing on various programs, and illustrated appropriately on the right-hand page.

The thesis of a STOP unit must be carefully supported by facts, thoroughly and logically presented in the unit. Often the customer already knows the bidder's experience and capabilities, and thus may quickly see through an empty boast. It is important to address the Oh-Ohs and negate competitors' Ghosts. In the long run, it is better to face company experience squarely and discuss how any problems have been eliminated than attempt to hide shortcomings.

If the company's experience is a major problem, then the affected STOP units must be carefully written and edited by the technical and marketing specialists, so that the proposal does not lose points in the scoring. Changes in technical approach and personnel should be emphasized, and the discussion should make clear how these changes have solved or will solve the problem.

Developing the STOP Profile

Once the outline of a proposal volume has been written and writers have been assigned for the sections, a STOP profile (a plan for the STOP unit) should be prepared for each section. A typical STOP profile form is illustrated in the Martin Marietta brochure in the Appendix. To prepare a STOP profile for review by the proposal manager and others in a storyboard conference, the writer of the unit must do the following:

- Understand the win strategies or marketing messages to be developed in the STOP unit.

- Understand what information (if any) in the unit will be classified. Classified information should not be included in the STOP profile. Keeping the STOP profiles unclassified will make it much easier for the entire proposal team to review the profiles in a storyboard or other planning conference.

- Write the section and subsection numbers and titles in the spaces marked "Section" and "Subsection" on the profile sheet. These numbers and titles should correspond exactly with the numbers and titles on the outline for the proposal.

- State the main idea of the unit clearly in the space labeled "Thesis." This thesis should be an idea, not a statement of details. It should clearly support at least one of the win themes of the proposal.

- Write in complete sentences the three or four main points that demonstrate the thesis.

- Sketch the illustrations to be used on the right-hand page, indicating clearly the content of each. Try to include the most appropriate illustrations for the content of the section, and try to vary the illustrations. A unit with different types of illustrations is much more effective than a unit with only one type of illustration. If photographs will be used, tape copies of them to the profile sheet. Sketch curves for line art, and provide column and row headings for tables, but write "classified" in place of any specific information that would make the illustration classified.

- Under each of the main points, list the details that will be presented to support that point. Be specific, but again write "classified" in place of any classified information.

For a successful proposal effort, it is essential that all STOP profiles be carefully prepared in time for the proposal manager's review or the storyboard conference. Reviewers need to be able to examine each section of the proposal, in order to:

- Ensure that all sections comply with the RFP (or the company's plan for an unsolicited proposal);

- Make certain that win strategies will be developed fully; and

- Assemble the best possible evidence to support the company's claims.

Usually the bid decision will have been made before this review, but company managers should review the STOP profile package to make sure that the company's proposal looks strong enough to warrant the investment in preparing the bid.

During the review of STOP profiles, the proposal manager and other technical and marketing experts should write very specific comments as needed on the STOP profiles, indicating material to be added or deleted, sources of additional information, *et cetera*.

The proposal will be stronger if the entire proposal team reviews the profiles, so that all writers and managers get a sense of how the proposal will be developed and have an opportunity to suggest other details and sources. For example, an engineer who has worked on a related program may be able to suggest test results that would strengthen a discussion of the company's previous experience with the technology.

Drafting the STOP Unit

While the review is in progress, the writers should be gathering materials and writing a draft of each STOP unit, paying careful attention to any page budgets and proving the thesis of each unit in both words and illustrations. If a writer is uncertain of what kinds of illustrations would best communicate the message, he or she should consult the volume editor, who is trained in strategies for presenting information.

When the writer receives comments from the storyboard, he or she should address them in the first draft, making it as specific, clear, and thorough as possible.

The writer of the first draft of a STOP unit should not be concerned with spelling, punctuation, or grammar. He or she should follow the proposal writing guidelines distributed by the proposal editor or manager, concentrating on accuracy, clarity, and thorough support for the thesis. The unit should be prepared neatly so that the text can be typed quickly. Also, each paragraph or illustration containing classified information should be properly labeled by the writer. Classified material omitted from the STOP profile should be added to the text and illustrations for the first draft of the STOP unit.

Writing a section of a proposal is much easier when STOP format is used, because the proposal or proposal volume will be broken into two-page or four-page units that are focused on separate main ideas. The thesis for the unit helps the writer to shape and control the discussion, and comments on the STOP profile (from the storyboard conference) provide additional information for the unit.

When a writer does have trouble writing a STOP unit, this difficulty will be apparent from the STOP profile, and the proposal manager can identify additional sources of information for the STOP unit, including other company personnel who might help.

The key to a good proposal written in STOP format is the storyboard conference or other review of the STOP profiles. The proposal manager must make certain that all supervisors and writers cooperate to prepare a good STOP profile for each STOP unit of the proposal.

Solicited proposals are usually prepared under a tight schedule. Because time is critical, any writer who cannot produce a good STOP profile for the storyboard or a good first draft of the STOP unit on time should be replaced as soon as possible. If the writer must stay on the proposal, the proposal manager must negotiate for the writer's time or cooperation.

10.2 STOP EDITING

A proposal editor's work should begin long before the RFP is released. With the arrival of the RFP, however, the editor's duties change, and he or she begins work on writing or editing an outline and a compliance matrix for the proposal.

The proposal team needs an outline of the proposal to use as a plan for the storyboard conference. The proposal editor should make certain that the outline of the proposal is responsive to the RFP (for a solicited proposal) or the company's plan. Every section of the outline should be divided into two- or four-page STOP units.

Also, the editor of a STOP proposal should write or edit a compliance matrix for the proposal, which can be a simple two-column list. One column identifies each section of the RFP that calls for discussion, and the other column lists the section of the proposal that provides that discussion. The compliance matrix is very useful to the proposal manager, writers, and reviewers of drafts, and is often included in the final submittal as an aid for those who will score the proposal.

The proposal editor sometimes reviews STOP profiles before they are submitted to the storyboard or other review conference, in order to assist writers with concepts for illustrations. But once the STOP profiles are written, he or she should always review them carefully:

- To understand more fully the proposal or proposal volume;

- To determine compliance with the RFP;

- To examine illustration concepts for appropriateness and balance;

- To check for logic and completeness in the development of each unit; and

- To identify units that need fuller development.

The Initial Review

Editing a technical proposal prepared in STOP format is easier than editing a proposal prepared in another format because each STOP unit is clearly defined in terms of subject, length, and organization. If the STOP units have been written carefully, then the editor has good material with which to work.

However, editing a STOP unit for a preliminary draft of a proposal is different from editing the unit for the final draft. To hold down proposal preparation expenses, the editor should edit preliminary drafts only lightly. Despite the fact that some reviewers will look for and point out typographical errors, misspellings, and so on in review drafts, the editor should not pay much attention to such problems. Rather, he or she should concentrate on the organization and clarity of each STOP unit, making sure that the words and illustrations fully demonstrate the main idea of the unit.

Because preliminary drafts of a proposal are reviewed for content, they need not be highly polished. Often units are revised radically after review, even if they have been written carefully and incorporate comments from the storyboard. So, the editor should not spend much time copy editing units for preliminary drafts. Nevertheless, the editor should take the following steps:

- Make sure section and illustration numbers are correct;

- Read the unit carefully to understand the content;

- Revise the thesis if necessary, so that it clearly and emphatically presents the marketing message of the unit;

- Examine the main points of the unit, to ensure that the thesis is developed directly and thoroughly;

- Examine the illustrations for variety, clarity, and proof of the thesis;

- Edit text very lightly and quickly, for format and readability; and

- Clean up illustrations quickly, for draft reproduction.

If the thesis is not clear, or the main points do not fully develop the thesis, then the editor should discuss the unit with the writer and clarify the thesis and main points. Similarly, if an illustration is not clear or is unnecessarily complex, the writer should be consulted on how it might best be revised. But fine points should not be debated or explored for a preliminary draft; fine tuning should be delayed until the unit has been reviewed in the first draft.

The more preliminary drafts of a proposal, the less quality in the first or second draft. A proposal or proposal volume that requires more than two preliminary drafts before the final draft is in trouble. Producing and reviewing too many drafts wastes too much time and energy, especially for the editor and others in the publications department. Careful staffing and a good storyboard conference can hold the number of preliminary drafts to the optimal one, or perhaps two.

Given the time and money it takes to produce final art for the proposal, whether using computer-assisted equipment or drawing on the boards, no illustrations should be released to the art department before first (or second) draft reviewers and the proposal manager are sure that the illustration will be used, unchanged, in the final draft of the proposal. The proposal editor, who is normally responsible for releasing art for production, should conserve money, time, and the energies of artists and proofreaders by observing this guideline.

The Final Review

For the final review draft of a proposal, the editor should edit each unit carefully. Revisions to the unit should be examined to ensure clarity and consistency with the rest of the unit and the volume. Text, line art, and tables should be edited according to standards distributed by the publications department lead for the proposal, and submitted to support groups for typing (or typesetting) and art production.

In final editing, the editor should focus especially on:

- Compliance with the RFP;

- Clarity and thoroughness of development of the thesis;

- Appropriateness of the discussion for the particular audience; and

- Correctness in content and presentation.

The editor should bring to the attention of the proposal manager any STOP unit that does not develop its thesis logically, thoroughly, and clearly.

The illustrations in the STOP unit must also be checked carefully, to make sure that they clearly and forcefully develop the thesis of the unit. The STOP format has been criticized as encouraging illustrations for the sake of having illustrations for right-hand pages. If the text of the unit is presenting solid evidence, however, there will be information in it that could and should be presented in illustrations for the STOP unit.

Usually, the editor's main problem with illustrations in technical proposals is not so much what to put in them as how to keep too much out of them. Illustrations that try to convey more information than the concept of the illustration will bear are difficult to interpret and lose their force. So each illustration, even large tables of data, should be designed and controlled so that the message of the illustration is clear. Each illustration should have a caption that states the message of the illustration.

The entire proposal must be written with the audience in mind. The editor must make sure that the material is presented effectively for the evaluators. The solicitation and award documents for contracting officers, executive summaries, technical volumes for technical experts, management volumes for management specialists, and cost volumes for finance specialists must be appropriate for the audiences. Solicitation and award documents and cost volumes are not prepared to STOP format, given their contents. But executive summaries and technical and management volumes are prepared to STOP format, and editing these different proposal documents effectively requires an understanding of the evaluators of each.

The Executive Summary. Executive summaries may be read by technical experts who want an overview of a large proposal. But summaries are read by a much broader audience than are technical volumes. Agency or company executives who are not technical experts will read the executive summary, so the technical content of this volume must be presented more simply. STOP format is particularly appropriate for this audience, because the content of the proposal is summarized in manageable, two-page units.

Executive summaries are often kept unclassified, and are made into attractive volumes that a contracting officer or executive might keep on his or her desk to show people the sort of work his or her contractors produce. Executive summaries are often printed in full color and usually have more illustrations, especially photographs, than the other proposal volumes have.

However, editors producing such a volume need to keep in mind the Federal Acquisition Regulation, which prohibits unnecessarily elaborate proposals. Proposal editors should try to produce an executive summary suggesting that the company does solid work but does not spend customers' money on elaborate productions.

The Technical Volume. Technical volumes are read primarily by technical experts, who score the proposal on the quality of the supporting data. Such volumes should not be ornate, but thorough and clear with technical discussions carried out in considerable detail. The greatest problem for the editor of STOP units for a technical volume is to make sure that the win strategies stand out from the mass of supporting evidence in the proposal, both in text and illustrations.

The Management Volume. Management volumes require clear organization by subject, so that evaluators can easily find the particular sections they seek. Certain sections, such as previous experience and key personnel, may be read by all the reviewers, but sections such as cost and schedule control, and facilities and equipment, may be read only by some reviewers. Organization charts should be detailed only to the appropriate level, and resumes should be edited so that the related experience of the proposed staff is emphasized.

Because management volumes present company policy and philosophy, writers tend to use "motherhood" descriptions (general, unsubstantiated claims about the company's expertise). The editor of a management volume must be more concerned than the editor of a technical volume about the specific evidence supporting the theses in the STOP units. Also, the editor must make sure that boilerplate is revised to support the particular win strategies of the proposal.

Whatever the volume, the editor of the proposal must make sure that the proposal is fully responsive to the RFP or RFQ. He or she must study the compliance matrix for the proposal carefully in order to be certain that all the discussion called for in the RFP has been provided in the proposal. Any lack of compliance should be brought to the immediate attention of the proposal manager, because failure to comply fully will cost the company points in the scoring of the proposal.

The proposal editor is often responsible for scheduling and coordinating all the activities of the publications department, including art preparation, typing or typesetting, layout, page make-up, proofreading, printing, checking pages, and binding. These demanding responsibilities are important to a successful effort. How to meet them is beyond the scope of this discussion, but it is important to remember that the proposal editor is usually the one charged with them.

10.3 PRODUCTION, LAYOUT, AND PUBLICATION

The technical or managerial experts who prepare technical proposals are often unaware of what the different groups in a publications department do to produce a technical proposal. All the members of a proposal team should be aware of the basic procedures in typing or typesetting, artwork, layout, paste-up, printing, and binding. These activities are very important to an attractive and technically sound proposal. They take time to do well—much more time than many people think.

Once the writers have produced the text and rough illustrations, and the editor has reviewed the work, the text goes to typists (and perhaps then to typesetters), and the art goes to the graphic artists.

Typesetting

Typists or typesetters put handwritten or typed material on disk or tape. The increased capabilities of word processing equipment have been a blessing and a curse for typists. Revisions to typed material are far more easily made; entire pages need not be retyped. But because changes are more easily made, typists are asked to make too many changes, too often, and too quickly.

Because the latest machinery is complicated, typing or typesetting often takes far longer than proposal managers anticipate. To expedite typing, the publications lead and the proposal manager must establish how many drafts of the proposal will be produced and how "clean" (free of handwritten corrections) the text should be for drafts, and typists should be given clean inputs.

Graphic Art

Graphic artists use manual techniques or computerized equipment to produce sketches for drafts of the proposal, and finished art for the final draft. Having detailed instructions from the editor greatly increases the accuracy of the artists' work and decreases the time required to produce the artwork. But it takes much longer to produce finished art than most people think, especially if computerized equipment is used to produce drawings that must be developed photographically, and if the art is complicated (scenarios, complex flow or organization charts, isometrics, and so on). Computerized equipment allows us to generate a new original of a piece more easily, but minor changes to a piece can take far longer on automated equipment than on the boards.

Before art is submitted to the graphic arts group, the group should be informed of the art standards for the proposal (sizing, line thickness, typefaces, *et cetera*). The artists should also know the approximate number of difficult, average, or easy pieces (judged according to publications department standards). Classified artwork may take longer to prepare because the artist and equipment must have the proper clearance.

Proofreading

Proofreaders' duties vary according to the material. For preliminary drafts, proofreaders should check typed text and drawn art against the originals, to make certain that nothing has been omitted or changed. In later drafts, proofreaders also check for errors or inconsistencies in spelling, grammar, punctuation, and sentence structure, and for errors or inconsistencies in content.

Proofreading is very demanding work. It requires great attention to details and a highly developed ability to spot problems. The time required for proofing text or art against the originals depends on the clarity of the originals. Proofing the corrections to the text or art does not consume nearly as much time, but accuracy becomes all the more important. Evaluators see mistakes as signs of ignorance or sloppiness.

Layout

Layout, which is the arrangement of words and art on a page, must be considered very early in the proposal preparation process, especially for proposals that are page--limited. When the writers prepare STOP unit profiles for the storyboard conference, they should visualize the space that their illustrations will take on the printed page and plan their illustrations for the unit accordingly.

From the first draft on, the proposal editor should "size" the text for the unit. While reduction of type is possible for the proposal, the text should be no smaller than eight point (if typeset) or 10 point (if typed), and letters in illustrations should never be smaller than six point (preferably eight point). The goal is readability, not volume of material, so reduction should not be used to solve page-limit problems.

Page Make-up

Page make-up (pasting or taping of text and art on a sheet of paper) can be done two ways: by hand or electronically. In the traditional manual method, illustrations and pieces of typed or typeset text are cut out, arranged on a mat (a sheet of paper that will be put on the printing press), and taped or glued down. Proper security or disclosure markings are affixed to each page. After these made-up pages are reviewed, they are numbered.

In the electronic method, pieces of text and art are manipulated in computer-assisted equipment, and each made-up, properly marked page is printed out. New originals for pages made electronically can be generated easily, but minor changes can take a long time to make, and they can be made only by a trained operator. Last minute corrections or changes cannot be made nearly as easily with automated equipment.

Great care must be taken with mats made up by hand; it may be difficult to generate a replacement mat, especially one containing artwork. But minor changes can be made more easily to pages made up manually, and many people are capable of making these changes.

Printing

The time required to print the copies of the proposal depends on the printing methods used and the capabilities of the print shop. It also depends on whether the pages will be printed on only one side of each sheet of paper, or on front and back.

Full-color (or "four-color") work is the most complex, time-consuming, and costly. It is generally used only for covers, for the executive summary of a large proposal, or for a very small proposal. If a job is done in full color, the entire job should be printed at the same time to ensure consistency in the colors. Four-color printing requires more skill than printing by other methods, and most print shops have far more one-color or two-color presses. It is very difficult to estimate how long four-color work will take. The printer must examine the profiles and an early draft of the proposal, and consider other work scheduled, to estimate how long the job will take. Given these complications, it is unwise to plan a full-color classified proposal.

If the proposal will be printed with the traditional black ink on white paper (a one-color job), then we can use various methods. Conventional office copiers can provide good quality reproduction. But photo-offset printing should be used for pages with photographs or screens (plastic overlays that will produce shaded areas on the printed page). Offset printing can be used for all the proposal pages, but it takes longer and costs more than photocopying, because of the differences in the processes. Whether photocopying, offset printing, or some other method is used, equipment and operators may have to gain clearance (if they have not already) to print classified material.

Binding

Binding copies of a technical proposal can be a very time-consuming process. Once the pages have been collated, each copy of each volume is bound. Some methods of binding, such as perfect binding, are not appropriate for proposals. But many choices are available, each having advantages and disadvantages.

© 1996 by Holbrook & Kellogg, Inc.

Saddle Stitching. In this process, the cover and the pages making up an 8 ½ by 11 inch volume are actually on 11 by 17 inch sheets, which are folded once to form an 8 ½ by 11 inch book, and stapled at the fold. The pages are held securely. The copy can be opened flat. But the thickness of the volume is limited, and it is very difficult to insert or remove sheets or sections.

Stapling. The 8 ½ by 11 inch pages (including covers) are stapled either once in the upper left corner or several times down the left side. The pages are held securely. However, stapled proposals have some disadvantages: the thickness of the volumes is limited, it is very difficult to insert or remove sheets or sections, and the volume does not open flat.

Looseleaf Binding. Pages of each copy are drilled with holes so that they fit a three- or five-ring binder. A proposal bound this way will open flat, sheets or sections can be inserted or removed easily, and binding is quick. Unfortunately, pages can tear at the holes and fall out, and binders of the proper size must be obtained. Also, it is often difficult to provide a cover for the binder.

Spiral Binding. A row of slots is punched along the left side of the cover and text pages, and a plastic spine with a "finger" for each slot is attached by inserting each finger through the appropriate slot in the pages. Sheets or sections can be inserted or removed relatively simply, by hand or machine, and printed covers can be used. Pages do not tear at the slots, and the volume opens flat. But hand binding is required, and care must be taken that the plastic binder is large enough that the proposal opens easily.

Whatever the type of binding, the pages for each copy of each volume of the proposal must be collated. Collating can be done by machine, if all the pages are one standard size; oversize or undersized pages must be interleaved by hand. Once a sample bound copy of each volume has been checked for accuracy of the collating, each copy of the volume can be bound. In a classified proposal, each copy of each bound classified volume must then be checked, page by page, to make certain that no pages are missing or duplicated.

The importance of a well-produced proposal cannot be overemphasized. If an evaluator receives a faulty copy, he or she will suspect incompetence in the bidder's organization. Doing production work perfectly takes time, and making sure that the production work has been done perfectly takes time also. Therefore, production time must be built into the proposal schedule, because the quality of the proposal will suggest to the evaluators the quality of the work that bidder would do on the contract.

10.4 QUALITY CONTROL CHECKS

Quality control checks for the proposal should take place throughout the preparation of the proposal. A successful proposal effort should involve a series of quality control checks of the STOP profiles or other planning materials, the preliminary drafts, and the final draft.

The storyboard conference is a good place to check the organization of the proposal and to check the arguments that support the win themes. The review of each draft of the proposal should be a quality control check; each reviewer should do what he or she can to improve the proposal. The proposal manager, reviewers, and editor should check each draft of the proposal for technical accuracy and completeness.

Throughout proposal preparation, the proposal manager and writers need specific information from reviewers. The author of a draft of the executive summary for a major proposal once wrote on the draft, "Comments are appreciated but are not too helpful. If something is wrong, fix it. If something needs to be added, add it."

Proofreaders should check each draft for perfection.

The proposal manager, writer, and editor should check the pasted-up mats of the gold copy of the proposal for errors in the final text and illustration corrections, and for proper classification markings. It is especially important to check captions for the illustrations and tables. In manual paste-up, the captions are usually prepared separately from the illustrations; thus, it is very easy for an illustration to have the wrong caption. Also, the editor should ensure that the quality of the mats is appropriate for the printing method to be used.

Before the copies of the proposal are bound, the editor should check a sample to ensure that all the pages reproduced well and were collated properly. Only then should the remaining copies be bound.

Once the proposal copies have been bound, a final quality control check is necessary, especially for a proposal containing classified materials. Each copy should be checked, page-by-page, to make certain that it contains every page and no duplicates, and that each page reproduced well. Although this task is laborious and time-consuming, it should be done. For a classified proposal, it must be done to avoid security problems.

The final check of the proposal should be done only by the proposal manager and the editor, with perhaps the assistance of the print shop lead. The writers should not be allowed to participate in this check, lest they try to make last-minute changes or impede a process operating on a very tight schedule. Anyone who missed the review of the final draft of the proposal should not take part in the final check, for the same reason. Changes at this point often threaten the schedule, and no one should be allowed to jeopardize meeting the deadline.

The appropriate time for the writers and others to review a printed, bound copy of the proposal is after the proposal has been shipped, not before.

Following these guidelines prevents ineffective, wasteful reviews, improper requests for revisions, and other unproductive steps—problems that make it difficult to adhere to the proposal schedule. Because it takes much longer to produce the printed proposal than most people think, there is always little time left at the end of the proposal schedule. The time that is left should be used for a carefully managed quality control check to make sure every copy of the proposal to be submitted is as good as it can be.

10.5 PACKAGING AND DELIVERY

Packing the deliverable copies of a proposal seems a mundane matter, but it is an important step in submitting a proposal. Certain steps should be followed, because it is important that the copies of the proposal arrive in good condition. It makes no sense to spend time and money producing an attractive proposal only to have it folded, torn, or stained in shipping.

Early in the proposal preparation process, the packing and delivery requirements should be established, in terms of the:

- Number of volumes;

- Length of each volume;

- Number of copies to be submitted;

- Classification or sensitivity of each volume;

- Packing date and time; and

- Delivery place, date, and time.

Proper packing materials should be ordered, and packing and marking assistance should be scheduled as soon as the requirements are known. If such packing is to be done in-house, the requirements should be discussed with the print shop, mail room, or other group in charge of packing and shipping.

Oversize materials such as engineering drawings may require reinforced cardboard shipping tubes of the proper size. These should be ordered as early in the proposal process as possible.

Packing an Unclassified, Nonsensitive Document

If only a few copies of a small 8 ½ by 11 inch document are being submitted, then stiff cardboard sheets should be inserted between each copy and inside the mailing envelope. The envelope should be securely taped and properly labeled with addresses inside and out.

If the submittal involves a number of copies of bound volumes, then a reinforced carton should be used. Each set of copies of each volume should be kept separate from the other sets. Cardboard sheets should be used to protect the covers of the copies facing the walls of the carton. The carton should be just large enough to hold all the documents; any extra space should be stuffed with packing materials so that the contents of the carton will not shift in handling. Again, the package should be properly labeled with addresses inside and out, and securely taped.

If the copies of the proposal volumes are stapled rather than spiral or looseleaf bound, then shrink-wrapping each copy will protect its appearance. However, shrink-wrapping can produce an ugly package, and our aim is to make the submittal attractive.

Packing a Classified or Sensitive Document

If the proposal contains classified or sensitive materials, additional labeling and wrapping will be required.

The packing and marking standards of the company and customer should be followed carefully. If we have any question concerning requirements, we should contact the company's security officer and the contracts staff assigned to the project. Before the submittal is wrapped and packed, we must fill out classified document control or transfer forms, using the copy numbers stamped on the covers of each copy of each classified volume, reference numbers such as the contract number for the procurement, and other identifying information. This paperwork must be filled out exactly. Even though little time is left in the proposal schedule, we must not rush this procedure, lest our mistakes result in security violations.

Delivery

Proposals should not be dropped in the mail, even if they are small submittals of unclassified or nonsensitive materials. For many different reasons, shipments can be delayed and deadlines missed. Package delivery services have failed to get packages to their destinations on schedule.

The safest means of delivery for a major proposal is to carry it by hand, whether the submittal is classified or not. If the proposal is traveling some distance, it should be carried or placed in baggage *on a regular, scheduled airline flight with a major carrier.* This precaution adds expense to the proposal effort, but it is far better to spend a few hundred dollars on an airline ticket than to risk having a proposal disqualified because it did not reach its destination in time.

On major proposals, it is wise to have a duplicate set of the submittal packaged and ready to ship in case something goes wrong with the original shipment. This may seem unnecessary, but it is far easier to forestall disaster than to recover from it.

Proposals often reach the customer on the actual deadline dates. Although this is clearly acceptable, it is advantageous to submit a proposal a day or two in advance of the deadline, to avoid the appearance of haste, even if the proposal ends up sitting in the bid room until the due date. Most customers would rather contract with a firm that plans its activities carefully and achieves quality in its work on a schedule that has time to spare. How a proposal is delivered can suggest a bidder's way of doing business, just as the proposal itself does.

10.6 ARCHIVING, CLEANING UP, AND DEBRIEFING

Everyone breathes a sigh of relief after the shipping of a proposal. But strangely enough, many people have difficulty adjusting to the sudden absence of the pressure of the final stages of the proposal. During this period immediately after the proposal is submitted, there is still proposal work to be done. How well it is done can affect the quality of proposals the company prepares in the future.

Archiving—the cataloguing and storing of materials used to prepare the proposal—is a very important activity after a proposal has been completed. Abstracts of the proposal should be prepared for the company's library or information center. Copies of the final proposal should be indexed and catalogued. Original artwork and photographs should be labeled with the name, date, and reference number of the proposal, and prepared for storage. The proposal manager or editor should work with company information specialists to ensure that materials are properly catalogued and indexed.

Not all materials should be archived, however. Copies of preliminary drafts should be disposed of properly, as should any working papers. (This is especially important for classified documents.) Borrowed materials such as specifications, standards, and old proposals should be returned promptly, to ensure compliance with future requests for materials.

Although a decision on the proposal may not be announced for months, the proposal manager should assemble the members of the proposal team for a debriefing on the proposal. Because proposal work is usually grueling, he or she should thank all the team members for their efforts. For those whose efforts went beyond the call of duty, the manager should prepare a report detailing their contributions for their personnel files.

Not every proposal is a winner. When one is not successful, the proposal manager should tell the proposal team what the evaluators reported. No manager likes to fail, but he or she should explain what went wrong without trying to assign blame to any specific person. When employees know what went wrong, it is easier for them to make everything go right the next time.

A formal debriefing on a successful proposal is a much more pleasant occasion. But some proposal managers go about such a debriefing the wrong way, by including only some members of the proposal team. Quite often, in large companies especially, many employees feel that their efforts on a successful proposal effort are not sufficiently recognized or rewarded, and that the credit for the win is given only to supervisors and managers. In some organizations this may be true, but a company that plans to prepare future proposals will encourage proposal managers to include the entire proposal team in the formal debriefing.

Everyone who contributed to the success of the proposal should be debriefed at the same time. This way, everyone has a sense that the proposal manager valued his or her efforts. As a result, everyone will be more committed to the success of the next proposal.

The proposal's success should also be celebrated with a party held off-site, after hours. At that social gathering, the proposal team can congratulate themselves on a job well done and reinforce the camaraderie that successful proposal efforts require.

CHAPTER 11

Proposals in Retrospect

11.1 THE TEAMS—WHITE, BLUE, RED, AND GRAY

Let's review, in proposal jargon, the color codes typically used to review the entire proposal process. Emotional tidal waves propel a proposal through the various stages, from pre-RFP marketing to the Gray Team's postmortem or victory review. Each aspect of the proposal should be charged with enthusiasm about how improvements will be made and about the prospect of winning. Each team approaches the proposal from its own unique perspective, and each is bent on asserting certain perspectives. Typically, the different attitudes of every team have to be blended, merged, and tempered into a smooth, persuasive argument.

The **White Team** members write the proposal and make these assertions:

- "We know the client and the RFP better than anyone."

- "Our work has more depth than any competitors'."

- "Our work is sacrosanct; hands off, please."

- "Management is on our side; the other teams really work for us."

The **Blue Team** tells the White Team that the competition is indeed formidable:

- "The competitors are better than you suspect; here is what they can do."

- "Here is an ideal firm—let's see your strategy to beat it."

- "Management is afraid you guys will be bowled over by the XYZ firm's steamroller."

- "Some of your concepts, plans, and themes leave something to be desired."

The **Red Team** further rebuts the White Team's cocky assertions:

- "What have you been doing for the past eight weeks?"

- "This is all wrong; you are 75 percent off target."

- "Did you read the RFP?"

- "Did you listen to the Blue Team?"

- "Wait until these proposal costs catch up to you!"

- "A massive rewrite is called for, today."

- "How come this is a 9 a.m.-5 p.m. operation? Where is your dedication?"

- "This is beatable stuff, really beatable stuff!"

Finally, the **Gray Team** analyzes the entire proposal process after someone has won the contract:

- "How can we repeat this victory?"

- "Did we make any mistakes?"

- "Did we make the short list, and best and final evaluations, If no, why not?"

- "Did we do a cost-effective proposal?"

In summary, we have a four-dimensional situation with which to deal. (See Figure 11.1 for a view of all teams working together.)

© 1996 by Holbrook & Kellogg, Inc.

THE WRITING (WHITE) TEAM
• Overall responsibility throughout, in charge of completion.
BLUE TEAM (Early in Proposal)
• Devises ideal competitors based on business intelligence. • Develops Straw Man concepts. • Analyzes the competition, its strengths and weaknesses.
RED TEAM (Mid Point)
• Plays the Devil's Advocate. • Suggests enhancements. • Checks methodically that the Writing (White) Team complies with each and every aspect of the RFP.
GRAY TEAM (Aftermath)
• Makes Post Mortem analysis, aftermath assessments. • Makes long range recommendations for improvements.
Note: If the Writing (White) Team is overwhelmed at any junction in the proposal process, it is standard industry practice to move Blue or Red team members over to supplement the writing team, and become members of the White Team until their part is complete.

Figure 11.1
The Four Teams for STOP Proposal Writing

11.2 ONE DIMENSION—THE WHITE (WRITING) TEAM

The White, or writing, Team is responsible for the overall management, writing, editing, and production of each stage of the proposal, as well as the final product. The writing team is under the specific direction of the proposal manager, who is also in charge of how the proposal uses resources from the Blue and Red Teams. The White Team is at the center of the effort, and all Blue, Red, and Gray Team activities ensure that the writing team does the following:

• Makes a schedule for the writing plan;

- Contributes text and art to the storyboard in a timely manner;

- Makes use of Ah-Has, Oh-Ohs, Ghost Stories, and Discriminators;

- Analyzes the RFP;

- Obeys the proposal directives; and

- Obeys the proposal manager.

The writing team has an enormous amount of work to do in 45 to 60 days, and no spare time. In smaller firms, the team usually wears a number of "hats," and is heavily burdened each day with two or three separate jobs on the proposal. At the larger firms, more staff members tackle the problems, but the scope of work and complexity of the proposal are also more demanding. The end result of being on the writing team is that a single-mindedness sets in, typically called tunnel vision.

This one-sidedness causes the staff to obscure minority opinion, blur finer distinctions, and press ahead relentlessly; this focused approach works well provided that no problems arise due to incomplete or faulty information. Prior to the advent of Red and Blue Teams by the aerospace industry in the 1960s, senior managers had to race through nearly complete proposals, and try to prove them against the RFP, client preferences, and cost strategy in time to meet the deadlines. This arrangement soured most people who had to try and make it work. Yet the writing team has to shoulder the workload and has to be the responsible party. Problems the writing team always faces include the following:

- Time is everybody's enemy. The 30 or 45 days from RFP release until the proposal deadline are never enough to ensure that every important task receives full approval from all members. The team works six days a week, 10 to 12 hours per day. There is no room for goof-offs.

- Things change rapidly for the team. Morale and discipline rise and fall. Feuds break out. Poor proposal managers are either replaced or supported by "helpers" from the Blue or Red Teams.

- The White Team feels a natural animosity toward the other teams, especially when the formal rebuttals of Blue or Red Team remarks necessitate taking time away from the proposal.

© 1996 by Holbrook & Kellogg, Inc.

- The team takes Blue and Red Team remarks personally, as though directly insulted.

The devotion and competence of the writing team are never generally challenged. The entire goal is to finish on time with quality. However, in order to break any monopoly on information management, the senior managers of most firms insist that the White Team cooperate, coordinate, and respond to Red, Blue, and Gray Teams.

11.3   TWO DIMENSIONS—THE BLUE TEAM

The Blue Team is upper management's friend, and is almost always the friend of the writing team. The Blue Team takes the time and energy to scour business development data, annual reports, capabilities studies, allies' reports, and general business knowledge to find out who the potential competitors are, what their strengths and weaknesses are, and how to tread more neatly through this mine field without allowing it to harm the proposal.

There are two significant products from the Blue Team:

- War game scenarios of how to cope with various individual competitors, or teams of competitors; and

- An ideal competitor, which consists of all the competition's best aspects—in other words, "if our proposal will beat this fellow's, it will beat anyone's."

Usually, a formal Blue Team briefing informs the White Team that some of its most cherished ideas are not so great. The Blue Team tells the White Team:

- "You may not be the perfect choice of the particular source selection board."

- "You overlooked a few things."

- "Lethal competitors are out there, who are armed with the best skills and talent money can buy."

The White Team digests specific, useful comments from the Blue Team and synthesizes them into the writing plan, usually without any hardship or animosity. After all, the Blue Team is helping to beef up the proposal's defenses. A Blue Team memo is usually written by the proposal manager, affirming or denying specific action items. Thus, the Blue Team gives friendly, new ideas, which fill out the one-sided thought process into a two-dimensional approach. We should now have the basic problems sorted out.

11.4 THREE DIMENSIONS—THE RED TEAM (THE DEVIL'S ADVOCATE)

The Red Team is not always welcomed by the writing team. Sometimes animosities and competition arise between them. The job of the Red Team is to criticize the proposal wherever needed, in as much depth and detail as they feel is productive. Typical Red Team remarks are:

- "We should not be in this sector of the market—abandon it now and let's cut our losses; cast a no bid."

- "The approach is 50 percent faulty."

- "The writing team rushed headlong into the effort and has produced an odd mixture of stuff."

- "The writing team's proposal is awful. We see no quality at all in the writing plan."

- "We, the Red Team, have a mandate from upper management to correct weaknesses (perceived and real), and make sure the writing team pays attention. The Red Team invokes this charter frequently, and assures management that the entire process will be beneficial." Figure 11.2 summarizes what to expect from a Red Team.

> THE RED TEAM WILL CHALLENGE:
>
> — ASSUMPTIONS
> — THEMES
> — APPROACHES
> — DATA
> — OPINIONS
> — MEANS OF PERSUASION
>
> "The only nice thing about being imperfect is the joy it brings to others."
>
> DOUG LARSON

Figure 11.2
Improving the Proposal

- "We take business development to task for overselling this RFP as opposed to other RFPs."

In these statements, the Red Team feels the White Team has been asleep at the throttle, and that little significant progress has been made. The Red Team is often correct. The White Team can be smug, complacent, and poorly advised. They ignored some of the key data in Figure 11.3. The members of the Red Team force those of the White Team to open their eyes. The storyboard enters into the fray in this manner:

- The Red Team manager needs to be fully competent to substantiate anything he or she says.

- The Red Team views blank spaces in the proposals as incompetence or ineptitude.

- The "light" sections awaiting later attention are harshly criticized.

- The overall progress and adherence to schedule are hit hard.

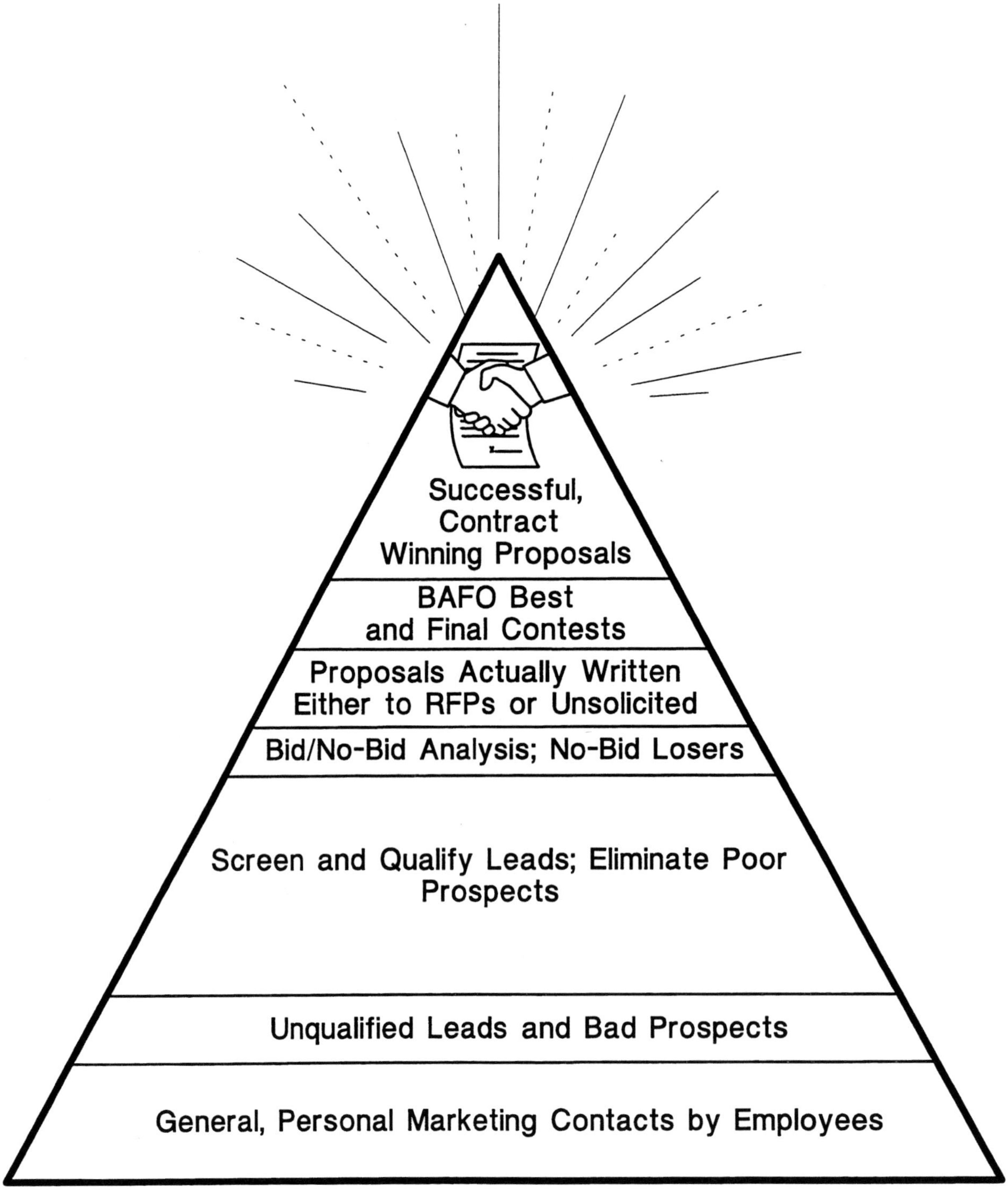
Successful,
Contract
Winning Proposals
BAFO Best
and Final Contests
Proposals Actually Written
Either to RFPs or Unsolicited
Bid/No-Bid Analysis; No-Bid Losers
Screen and Qualify Leads; Eliminate Poor
Prospects
Unqualified Leads and Bad Prospects
General, Personal Marketing Contacts by Employees

Figure 11.3

The Red Team's Review Reflects Its Knowledge of the Ratio of Successful Proposals to Overall Marketing Actions

© 1996 by Holbrook & Kellogg, Inc.

- The RFP's real and hypothetical intentions are questioned.

The Red Team imitates the source selection board's evaluation in its thoroughness in assessing the quality of the proposal. Table 6.3 shows examples of SSB checklists, which enable each Red Team member to score the proposal, and pass on the hypothetical findings to the writing team for validation. (Every Red Team should have a checklist like Table 6.3 for its uses. Tailoring the RFP will be useful.)

After receiving one or two agonizing Red Team documents, the writing team has a validated, systematic set of critiques that make the proposal three-dimensional. After the proposal manager, the writing team, and the Red Team have hashed out all corrections, the Red Team synthesis is complete, and the proposal's quality should have drastically improved.

The flashback phenomenon sometimes occurs, in which the writing team actually tries to overlook or ignore Red Team's remarks. No credible management structure will tolerate such a waste of energy and effort. Industry-wide, 25 to 50 percent of advice given by Red Teams is followed; the rest is dismissed for some good reason, usually in a written format that accounts for each single critique.

Massive rewrites emerge out of the Red Team's critique; and the writing team members see on the storyboard those areas requiring more work. No excuse for lateness is acceptable. The natural "us *versus* them" contentions between Writing and Red Teams must be well managed to ensure that the proposal is the ultimate beneficiary, and that all the rewriting is useful. If our firm does not have excellent Red Team members, we should hire consultants to fill the void.

In conclusion, Red Teams are the catalyst that forces White Teams to prove their contentions, and to write improved proposals. The length, width, and depth of the issues are now covered.

11.5 THE FOURTH AND FINAL STEP—THE GRAY TEAM (WIN-LOSS ASSESSMENT)

A proposal effort is not complete without a thorough accounting of why that effort won or lost. Without the documentation needed to identify strengths and weaknesses, technical publications groups are handicapped when they try to avoid past failures, or to repeat successes. By conducting a formal postmortem for each proposal, a technical publications Gray Team can enhance the company's archives, evaluate proposal costs, and objectively retain significant factors that contributed to a proposal's victory or defeat (see Section 2.7.1 for a review of this topic).

On a typical annual win-loss sheet, let us suppose a hypothetical firm proposed on 36 contracts; 12 major and 24 minor. The Gray Team reports the following:

Major	Minor
12 bid; six won (50%)	24 bid; nine won (37.5%)
10 of 12—best and finals (83%)	12 of 24—best and finals (50%)
2 outright losses (16%)	12 outright losses (50%)

Further tabulation is needed to identify why the outright losses occurred. It is logical that a contender will be in the best and finals category 75 to 100 percent of the time. Here is a scale of how proposals are typically scored in government evaluations:

- Without merit
- Acceptable with modification
- Acceptable
- Competitive range
- Best and final range
- Award winner

The hypothetical firm is troubled because two major and 12 minor proposals were in the "without merit" category, which embarrassed the company and wasted a lot of money. The most encouraging fact is that 50 percent of the large contracts were won, which is quite an accomplishment. The Gray Team praises the entire firm. A close examination of the losing proposals yields the following profiles:

Major Proposals

- We won one-half.

- We were finalists almost 85 percent of the time.

- Our losses were far from cataclysmic.

- We were on target all year long, and bid on the right jobs.

- Our revenue and reputation are good.

- Our clients are happy and receptive to new offers.

Minor Proposals

- We won over one-third, which is above the industry average.

- We reached best and final evaluations only half the time.

- Our prices cost us some of the close contests.

- Our technical approach was clearly unsatisfactory to many clients.

- The Red Team's remarks were omitted and negated; the Gray Team finds that the losers were equally careless.

- Three losses came from the same proposal manager.

Prognosis-Major Efforts

- We are pretty much on track.

- We cast "no bids" for bad, "wired" RFPs and selected only those which we could win.

- Our cost structure wins, too.

© 1996 by Holbrook & Kellogg, Inc.

- Next year, we will aim for more of the same.

Prognosis-Minor Efforts

- We are not doing too badly, and won some profitable jobs.

- We got to best and finals often, and lost because of cost.

- Some new clients did not turn out well.

- Some proposal managers did not meet our expectations.

- One-half of our losing proposals were poorly marketed; postmortems from the clients revealed that we were overly optimistic about our odds of winning.

- Some Red Teams were not persistent enough.

- Our inadequate knowledge of cost caused us to lose close decisions.

- Management had us pursue several losing efforts only to keep everybody busy between major proposals.

Next Year's Synthesis-Full-Cycle Planning

On major proposal efforts, we did well for the most part. Accordingly, we should continue with the winning pattern.

However, on minor proposal efforts, we need to modify our behavior. We will:

- Cast a "no bid" for RFPs done by clients who uniformly rejected this year's proposals;

- Cast a "no bid" on RFPs where cost outweighs technical advantages;

- Cast a "no bid" on RFPs where our management may use the effort as a filler;

- Demand better, more thorough marketing insight;

© 1996 by Holbrook & Kellogg, Inc.

- Select more industrious proposal managers;

- Keep close track of the nine new, smaller contracts; and

- Try to limit the number of minor proposals to two-thirds of last year's total; for example, do 18, not 24, smaller proposals, and focus more intensely on them in order to increase the win ratio.

We must not rob the large, more successful proposal efforts only to "share the wealth" with the smaller ones. Such a step will only water down our winnings for no appreciable purpose.

In addition, we must take our lumps from upper management and marketing people who feel we should win all the time; we must refute their arguments with statistical presentations of facts, year in and year out.

In the long run, winning proposals are produced by improving our procedures each time, and by correcting the errors identified in the closed-loop system. The Gray Team's official purpose should be the careful recording, evaluation, and sharing of win-loss information.

Other Publications from Holbrook & Kellogg

You may order any of these publications by writing or faxing to:
Holbrook & Kellogg, Inc. • 1964 Gallows Road, Suite 200 • Vienna, Virginia 22182
Fax: (703) 506-1948 • Phone: (703) 506-0600

NEWSLETTERS

Federal Acquisition Report
1 year—$114; 2 years—$199; 3 years—$285.
Monthly news and analysis of contracting issues.

Acquisition Issues
1 year—$299; 2 years—$565. Monthly in-depth analysis of a single contracting topic from the businessperson's perspective.

REGULATIONS

FAR 90 Update Service 2-volume looseleaf. Up-to-date FAR *plus* 1 year of FACs (updates) and monthly *FAR Alerts*—$227. One year of FACs only—$173.

DFARS Update Service 2-volume looseleaf. Up-to-date DFARS *plus* one year of updates (DACs)—$296. One year of DACs only—$109.

FIRMR Update Service Looseleaf. Copy of FIRMR, *plus* one year of update—$199.

Federal Travel Regulation Update Service
Looseleaf. Current FTR *plus* one year of updates—$110. Updates only—$62.

BOOKS

Charge It! Guide to the Government Purchase Card Program Looseleaf—$55. Covers the new simplified acquisition procedures that allow the use of the government credit card. Includes common pitfalls, program guidelines, and sample procedures.

The Complete Protest Desk Guide
Looseleaf—$139. A practical guide to protests before contracting officers, GAO, GSBCA, and the courts. Includes model protest letters and updates with a quarterly newsletter.

Complying with FASA: A Comprehensive Guide to the Federal Acquisition Streamlining Act
Looseleaf (2 vols.)—$250, *includes 4 quarterly updates*. A simple roadmap for understanding and dealing with the new federal procurement laws. Provides explicit and concise guidance on FASA provisions and their implementing regulations.

Desk Guide to Cost & Price Analysis
Looseleaf—$127. An exhaustive guide to cost and pricing issues ranging from policy and concepts to submission of cost or pricing data, from analysis of direct labor cost to profit determination.

Preparing for a DCAA Audit Package
Looseleaf—$129. Includes Holbrook & Kellogg's DCAA Accounting System Survey, DCAA Estimating System Survey, and Preparing for a DCAA Audit in one volume.

Desk Guide to Preparing Statements of Work
Looseleaf—$130. Describes the different types of SOW documents and when to use them. Includes many varied sample statements of work.

Drug-Free Workplace Program
Looseleaf—$98. Applies the policies and procedures that contractors need to implement to comply with federal Drug-Free Workplace regulations.

Fundamentals of Federal Contract Law
Hardcover—$60. Provides a thorough and detailed overview of the federal contract process as interpreted by federal courts, the Comptroller General, and boards of contract appeal.

How to Create and Present Successful Government Proposals Hardcover—$55. Reveals proven techniques for preparing effective proposals.

How to Write a Statement of Work
Looseleaf—$93. Untangles the information that goes into an SOW, lays out clear, well-organized format, and shows how each element relates to other portions of the solicitation and contract.

Juran's Quality Control Handbook, 4th Edition
Hardcover—$98. The perennial international desktop guide for continuous quality improvement, updated with new processes for managing quality planning, control, and improvement.

Multiple Award Schedules: Guide for Business and Government Looseleaf—$86. Covers recent changes to the regs governing schedule contracts. Provides guidance on how to win a schedule contract, how to sell from one, and how to respond to government audits.

Paperless Contracting: The EDI Revolution Looseleaf—$95. Lays out the government's EDI programs and progress. Gives tips for implementing electronic data interchange at your organization to take advantage of the increased opportunities and efficiencies of EDI in the contracting arena.

Past Performance in Government Contracting—A Desk Guide for Contractors and Agencies Looseleaf—$110. A team of industry and government experts put together this comprehensive text covering all areas of implementing past performance as an evaluation factor, including legal aspects, agency and contractor perspectives, impact on evaluations, and information/technology requirements.

Preparing and Delivering Effective Technical Presentations Softcover—$50. A how-to full of practical tips to help you address the specific problem of transferring technical knowledge to your audience.

Procurement Ethics Desktop Reference and Update Service, Revised Edition Looseleaf—$169. Spells out rights and responsibilities, as well as the letter of the law and the penalties for violating ethics standards in contracting. Includes quarterly updates.

Procurement Manager's Guide to EC/EDI Softcover—$96. Practical approaches for implementing electronic commerce in the procurement process. Includes policy issues as well as the technical and analytical tools to apply information technology to procurement management.

Quality: The Reference Book for Government Contracts Hardcover—$67. Demonstrates, step-by-step, how government agencies assure that contractors meet quality requirements.

8(a) Program Guide (SBA's Minority Small Business & Capital Ownership Development Program) Looseleaf—$69. Explains all the policies, procedures, instructions, and guidelines you need to manage your 8(a) status.

Selling to the U.S. Government Spiral—$49. Concise overview of the federal customer. Covers all areas of marketing, contract administration, and regulations. Great intro for the new contractor; excellent quick-reference for those with more experience.

Source Selection: A Seller's Perspective Softcover—$69. Will help you understand and persuade your customers, turn government requirements into sales opportunities, and sell systematically, consistently, and successfully.

Technical Evaluations in Government Contracting Softcover—$85. Focuses on the process of technical evaluation and how it affects the negotiation and award of contracts. Covers developing a plan, selecting evaluation factors, including past performance, writing proposal preparation instructions, and much more.

Terminations for Convenience: A Contractor's Guide Looseleaf—$99. Shows you how to recover the maximum amount of costs in a terminated contract. Strategies for surviving a "T for C" from responding to the notice of termination to organizing the termination team to negotiating the final settlement.

A User' Guide to Federal Architect-Engineer Contracts Hardcover—$120. Leads contractors through the A-E process, from marketing and preselection to responding to RFPs, from cost negotiations to reaching agreement.

Using the Freedom of Information Act as a Business Tool Softcover—$55. Shows business how to tap into the U.S. federal government information vault. Discusses what information is available and how to get it. Includes sample request letter, fee structures, and contact points.

Winning Negotiations in Federal Contracting Looseleaf—$83. Transforms general negotiating principles into practical techniques for winning government contracts.

Prices effective 4/1/96 and are subject to change.

Holbrook & Kellogg also offers on-site seminars and workshops in a wide range of federal contracting areas, tailored to meet your organization's individual needs.

Call (703) 506-0600 or toll-free (800) 506-4450 for more information.